GIANT SNAKES
unwravelling the coils of mystery

Michael Newton

Typeset by Jonathan Downes,
Cover and Layout by *F.citroiii* for CFZ Communications
Using Microsoft Word 2000, Microsoft , Publisher 2000, Adobe Photoshop CS.

Photographs © 2009 CFZ except where noted

First published in Great Britain by CFZ Press

CFZ Press
Myrtle Cottage
Woolsery
Bideford
North Devon
EX39 5QR

THE CENTRE FOR FORTEAN ZOOLOGY

www.cfz.org.uk

© CFZ MMIX

ISBN: 978-1-905723-39-3

To the memory of Percy Harrison Fawcett

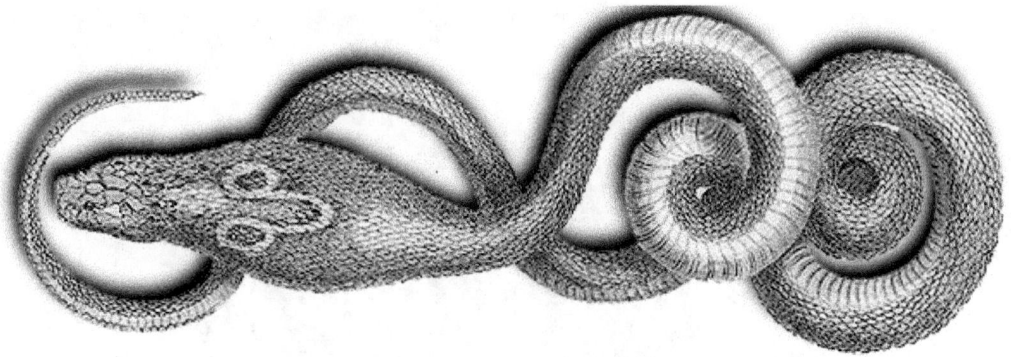

Contents

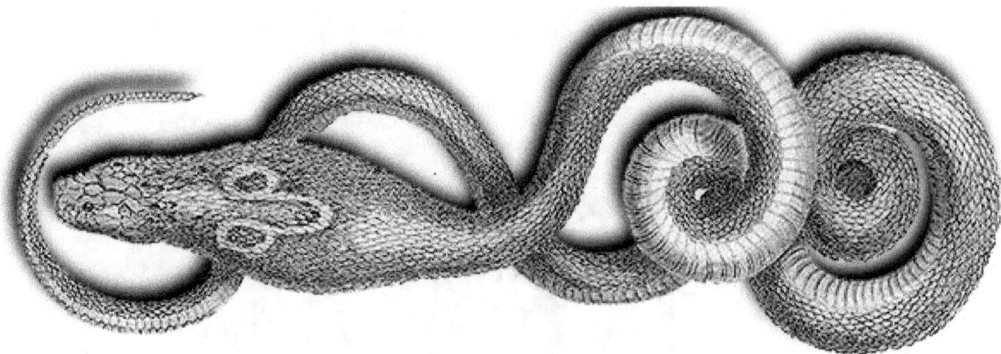

Preface

S nakes captured my imagination as a child. Perhaps it was the Tarzan films that seemed so common in those days, or the dramatic *Bomba the Jungle Boy* novels by `Roy Rockwood` that delivered deadly serpents with a clockwork regularity that never failed to thrill a budding would-be herpetologist. In high school I discovered *On the Track of Unknown Animals,* by Dr. Bernard Heuvelmans (the "father" of cryptozoology), and learned that snakes I viewed as huge might be mere infants next to some reported from the Amazon. More years would pass before I realized that Heuvelmans had barely scratched the surface with his tales of giant serpents prowling South America. In fact, it seemed that they were *everywhere,* some of them nearly in my own backyard.

To date no author has attempted to collect and publish all known data on reports of truly giant snakes around the world. This volume aims to fill that gap in scholarship and pave the way for future research on a fascinating subject. I owe a heartfelt debt of gratitude to Jonathan Downes and the Centre for Fortean Zoology for bringing it to print.

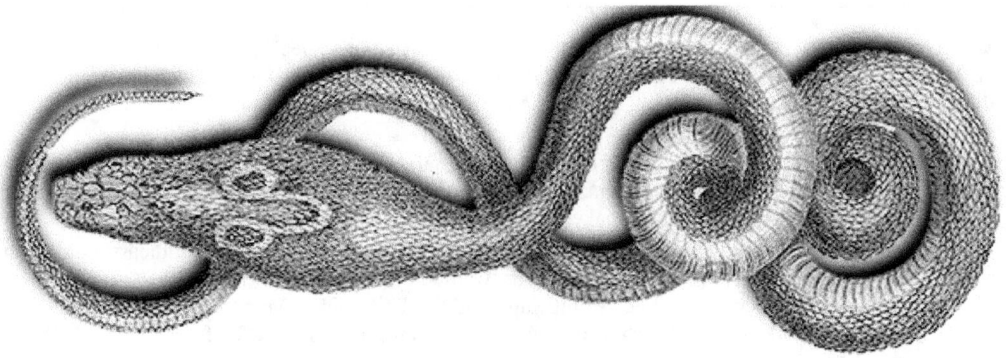

Chapter 1: The Big Six

L ove them or hate them, snakes elicit strong emotions from most human beings. Huge serpents infest religion and mythology: the Greek *Ladon* and Norse *Níðhöggr*, Egypt's *Apep* and Persia's *Zahhāk*, India's *Shesha* and Thailand's *Nāga*, the Mayan *Gukumatz* and Aztec *Quetzalcoatl*. Ouroboros, devouring its tail, comprised the Milky Way, while giant Mucalinda sheltered Buddha from a storm. Australia's aborigines revered the Rainbow Serpent, as voodoo initiates sacrificed to *Damballah-Wedo*. Satan took serpentine form in the Garden of Eden, whereas Moses used a brass serpent to heal snakebites.[1]

Giant snakes of indeterminate species are also a staple of horror films, beginning with *King Kong* in 1933. Inflated atom-bomb survivor Glenn Manning grappled with a huge snake in the wilds of Mexico, in *War of the Colossal Beast* (1958), while Dino DeLaurentis kept the giant snake but cut the dinosaurs from his remake of *King Kong* in 1976. Future California governor Arnold Schwarzenegger slew the giant house-pet of a snake cult's priest in *Conan the Barbarian* (1982) - and there the trail went cold until the advent of computer-graphic technology placed titanic beasts within the grasp of every film producer on Earth.[2]

The new trend in serpentine horror began with *Anaconda* (1997), merging live action with CG effects and presenting the last big-name cast to be seen in a giant-snake film. Then came the low-budget deluge, including *King Cobra* (1999), *Python* (2000), *Python 2* (2002), *Boa* (2002), *The Anasazi Mummy* (2002), and *Boa vs. Python* (2004). *Anacondas: Search for the Blood Orchid* (2004) unaccountably transplanted South American serpents to Malaysia for a tourist feeding-frenzy, while Philippine director Chaninton Muangsuwan gave us *Boa...Nguu Yak!* (2006) and Native American magic launched the rampage of *Mega Snake* (2007). As I prepared this manuscript for press, no details were available for *Anaconda 3: The Offspring* or *Anaconda 4: Trail of Blood,* slated for television broadcast in 2009.[3]

Of course, such monsters only live on-screen ... or, do they?

The strange and unsettling truth is that reports of monstrous serpents have been filed from every continent except Antarctica, nor are they limited to ancient times. This is their story, frequently as gruesome and fantastic as the wildest tale from Hollywood, but which - if true - challenges every major tenet held by modern herpetology.

When mainstream scientists refer to *giant snakes* they mean known species commonly described as the "Big Six."[4] While experts often disagree over the record size attained by any given species, most agree that these six are the largest snakes on Earth. The list includes:

- **The green anaconda** (*Eunectes murinus*), an aquatic boa inhabiting swamps and rivers of South America, most famously the Amazon and Orinoco Rivers and their tributaries. Some authorities consider it the world's largest snake and nearly all agree upon it as the heaviest.[5]
- **The reticulated or regal python** (*Python reticulatus*), found throughout Southeast Asia from the Nicobar Islands, Myanmar, Thailand, Laos, Cambodia, Vietnam and Malaysia, eastward through Indonesia, the Indo-Australian Archipelago and the Philippines. Many herpetologists consider it the longest snake on Earth, while granting that the anaconda is, foot-for-foot, substantially heavier.[6]
- **The African rock python** (*P. sebae*), found with two subspecies in grasslands and savannah throughout sub-Saharan Africa, universally granted status as one of the world's largest snakes.[7]
- **The Indian python** (*P. molurus*) shares much of its range with the larger reticulated python, inhabiting Pakistan, India, Sri Lanka, southern Nepal, Bangladesh, Myanmar, southern China, Thailand, Laos, Vietnam, Cambodia, Peninsular Malaysia and Indonesia. Its Burmese subspecies (*P. m. bivittatus*) commonly ranks fourth on Big Six rosters and has colonized the swamps of southern Florida.[8]
- **The amethystine python** (*Morelia amethistina*) is Australia's largest known snake, found in bushlands, rain forests, and suburbia. Its range also includes Papua New Guinea and certain Indonesian islands.[9]
- **The common boa** - often called the "boa constrictor," though *all* boas and pythons kill their prey by constriction - is another Latin American serpent, with 10 smaller subspecies scattered from Mexico to Argentina, plus colonies on the Caribbean islands of Dominica, St. Lucia, and Trinidad-Tobago. Hardly a giant beside its Big Six competitors, the common boas rarely exceeds 10 feet in length. The largest in captivity today is a 15-foot female from Suriname, displayed at California's San Diego Zoo. A supposed 18.5-foot specimen from Trinidad, reported by author James Oliver in 1963, proved to be a misidentified anaconda.[10]

And yet ....

As detailed in subsequent chapters, sightings of *much* larger snakes have been reported throughout history. The accounts include both anacondas and pythons, as well as apparent new species unknown to science. More startling still, those reports are not confined to dense jungles in Asia, Africa, or South America. They also emanate from Europe, from Great Britain, and from the United States.

In 1912 ex-president Theodore Roosevelt and the New York Zoological Society - which he co-founded in April 1895 - offered a $1,000 reward for the capture of any snake measuring 30 feet or longer. Over time, inflation boosted the reward to $10,000, then to its present $50,000, offered by the New York-based Wildlife Conservation Society. The terms of that offer require delivery of the serpent - alive and in good health, with all required permits and other paperwork - to the Bronx Zoo in New York City.[11]

So far, the cash remains unclaimed.

Snake *skins* exceeding 30 feet have been recorded on a number of occasions, but as every taxidermist knows, such trophies may stretch 30 percent or more during tanning. Thus, a 30-foot hide may represent a living snake no more than 20 feet in length. That said, while 20-footers may not win the money pledged by Teddy Roosevelt so long ago, they can be giants in their own right - especially if found in

North America or Europe, where official records for the largest known snakes fall short of nine feet.[12]

Proving the case for giant snakes without a corpse or living specimen is no easy task. Eyewitness evidence - the kind preferred above all else in courts of law, where lives and liberty hang in the balance - is automatically dismissed by self-styled skeptics and most mainstream scientists. They note, quite properly, that size of living serpents is notoriously difficult to estimate, particularly if the witness is surprised, frightened, or unfamiliar with known species in a given area.

Author Chad Arment, in his definitive account of North American giant-snake sightings logged between 1714 and 1960, describes two broad categories of erroneous reports. The first type - completely false stories - includes deliberate newspaper hoaxes (regrettably commonplace in the 19th century); fabrication of sightings, tracks, and other evidence by individual pranksters; misidentification of common objects (fire hoses, pipes, etc.) as snakes; lies told to frighten children or trespassers, as a means of social control; and slow evolution of fictional tales into "fact" or urban legends.[13]

In Arment's second category, simple ignorance or inexperience spawns tales of huge serpents when witnesses exaggerate the size of known snakes briefly glimpsed under less-than-ideal conditions. Compound that problem with sightings of "freak" specimens larger than average size, and the stage is set for another tall tale.[14]

A final joker in the deck is the appearance of exotic snakes at large throughout North America, Europe, and the United Kingdom. Various boas and pythons are popular pets, but they sometimes escape or outgrow their terrariums, whereupon some negligent owners release them into the wild. As a result, we know that Burmese pythons breed and thrive in southern Florida: 67 were killed between 1992 and 2004, while two others measuring 16 and 20 feet respectively were caught alive. Meanwhile, Kansas campers killed a 20-foot Burmese python in 2004, and a 19-footer was found dead in Indiana's White River two years later.[15]

Once the escapees and releases are eliminated, what remains?

Clearly, reports of giant snakes are either true or false. In judging their veracity, we must consider any factors that suggest a hoax. Tales rated trustworthy may still involve innocent misidentification of known species or unwitting exaggeration of a specimen's size. The final possibility - sightings of a new species unclassified by science - is the most difficult to prove (impossible, perhaps, without a specimen in hand) but still deserves consideration if the circumstances warrant it.

We shall begin our global search in the least likely venue for a giant reptile - Europe - and proceed from there around the world until we reach those steaming jungles where, as Sir Arthur Conan Doyle's Professor George Challenger once suggested, "nothing would surprise."

And we shall see, perhaps, that even the professor was mistaken.

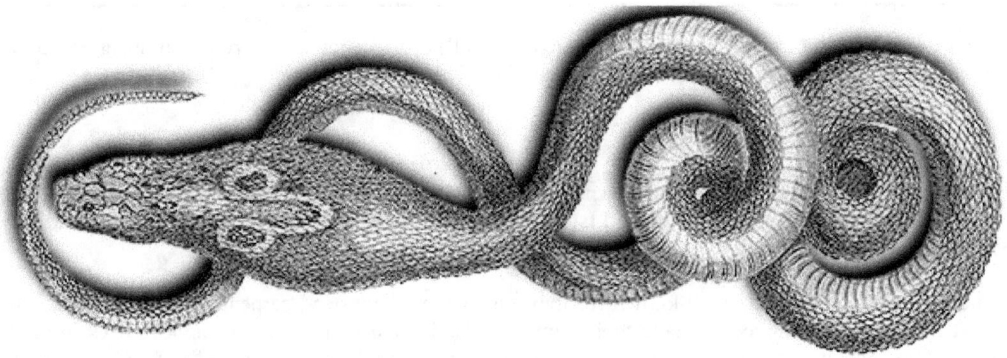

Chapter 2: Europe

Europe, at a glance, seems an unlikely venue for pursuit of giant snakes. Five European species known to science may exceed six feet in length, including the large whip snake (*Coluber jugularis*), two species of Aesculapian snake (*Elaphe longissima* and *Zamenis lineatus*), the blunt-nosed viper (*Macrovipera lebetina*), and the Montpelier snake (*Malpolon monspessulanus*). Ten other species boast record lengths exceeding three feet but stopping short of four. They include the western whip snake (*Coluber viridiflavus*), Balkan whip snake (*Hierophis gemonensis*), Dahl's whip snake (*Platyceps najadum*), leopard snake (*Elaphe situla*), sand boa (*Eryx jaculus*), coin snake (*Hemorrhois nummifer*), dice snake (*Natrix tessellata*), grass snake (*Natrix natrix*), European adder (*Vipera berus*), and European cat snake (*Telescopus fallax*).[1]

None of those reptiles qualify as giants, nor are Europe's 13 other species likely to be mistaken for slithering monsters. Still, reports of much larger serpents span the continent, ranging from classical antiquity to modern times. For clarity's sake, we shall survey the continent from west to east, one region at a time.

Ireland

Religious folk credit Saint Patrick with driving all snakes out of Ireland, sometime between 433 and 493 A.D. Modern herpetologists take a different view, maintaining that no snakes ever existed in Ireland, thanks to its geographical isolation from Britain and cold coastal waters.[2] That fact - if fact it is - has not, however, banished giant snakes from Irish lore.

Legendary hero Finn Mac Cumhail (A.K.A. Finn MacCool) reportedly slew a host of huge serpents throughout Ireland, during the 3rd century. His conquests included terrestrial serpents killed at Áth Cliath, Benn Eadair, Glenarm, Glen Inne (two in one battle), and Glenn Dorcha, as well as large aquatic snakes at Lough Cera, Lough Erne, Lough Foyle (another double-header), Lough Laeghaire, Lough Lurgan, Lough Mask, Lough Meilge, Lough Neagh, Lough Ramhuir, Lough Ree, Lough Riach, Lough Sileann, and the River Bann. In fact, we are told, "there was not a reptile in Ireland's glens but he took by the force of his blows." The final tally is uncertain, but Finn reportedly "exterminated every monster against which he advanced."[3]

Most historians and folklorists assume Finn's serial slaughter of serpents belongs solely to myth. Author

Peter Costello, by contrast, suggests that some (if not all) of the legends may represent exaggerated tales of encounters with real-life reptiles, presently unidentified.[4]

Great Britain

The grass snake is Britain's largest known reptile, with a record length of 3 feet 11 inches.[5] Nonetheless, reports of giant reptiles on the Sceptered Isle rival the most outlandish claims surrounding Ireland's Finn MacCool.

The Scottish highlands are well-known for their elusive lake monsters, serpentine and otherwise, but at least three legendary serpents made their homes on land - and still make rare appearances in modern times. One, a relatively modest monster, was the *beither*, which derives its name from a Gaelic word for "beast" or "serpent." While sometimes found in Highland lakes and rivers, the *beither* also favors caves. A Strathmore gamekeeper reported several sightings by his in-laws in the 1930s, around Loch a' Mhuillidh, near Glen Strathfarrar, and on the slopes of the mountain Squrr na Lapaich. More recently, in 1975, five witnesses reportedly saw a *beither* near the Falls of Kilmorack and tried to catch it, but the snake escaped into a gorge near Beaufort Castle.[6]

Author Richard Freeman rates the 12th-century Linton worm of Roxburghshire, Scotland, as "Britain's laziest dragon" - so lazy, in fact that it refused to chase its prey, preferring to lie hidden in a cave on Linton Hill and snatch humans or livestock passing by outside. When it had fed, the snake emerged from hiding and coiled around the hill, leaving deep impressions of its massive body. Local villagers placed a bounty on the snake's head, thus luring the Laird of Linton from his home in Somerville to kill it. This he accomplished with a spear dipped in boiling pitch, resin and sulfur, which he rammed into the reptile's gaping jaws. As his reward, the laird not only banked the peasants' cash but also won appointment as the King of Scotland's Royal Falconer.[7]

A large albino serpent once allegedly inhabited the Galloway district, growing large enough for its coils to encircle Mote Hill, near Daltry. As described by Scottish folklorist Andrew Land in 1885, the lord of Galloway offered rewards for the reptile's destruction, but it swallowed one of his knights and thus discouraged further attempts on its life. A local blacksmith built himself a suit of armor equipped with retractable spikes and let the serpent swallow him, then rolled about inside with spikes extended, freeing himself as he gutted the snake from within. Author Richard Freeman suggests that the Daltry serpent inspired Bram Stoker's novel *Lair of the White Worm* (1911).[8]

British folklore teems with dragons, including winged fire-breathing monsters that bear no resemblance to snakes, but Richard Freeman notes that *five* distinct types of dragons apparently terrorized England at various times. One species, the "worm" or "wyrm," possessed no legs or wings and belched no flames (although some specimens spat deadly venom, like some modern cobras). Worms were, in short, the very models of our giant snakes.[9] A brief alphabetical listing includes:

- **The Bignor worm**, a snake that mimicked the Daltry and Linton monsters, coiling around Bignor Hill, in Sussex, and marking the earth with its body.[10]
- The forest-dwelling **Bishop Auckland worm**, in Durham, which devoured man and beast alike until a member of the local Pollard family fought and killed it with a sword. Pollard gave his weapon to the Bishop of Durham, receiving a grant of land in return.[11]
- **The Brinsop worm** of Herefordshire, reportedly lured from its lair in a well and dispatched by Saint George. While this was not the saint's most famous reptile-slaying, a 12th-century carving inside Brinsop's church depicts George spearing a serpent as long as his horse.[12]

- **The Crowcombe worm**, from Somerset, described as a snake with a body thicker than the largest oak tree in Shervage Wood, where it dwelled. After preying on livestock, the reptile devoured two gypsies and a local shepherd. Subsequently, a woodcutter from nearby Stogumber met the beast, while picking berries in the forest, and hacked it in two with his axe.[13]

- **The Fittlesworth worm**, a relatively modern snake reported from that part of Sussex in 1867. Published accounts describe the reptile hissing and threatening humans who passed by its lair, but its fate is not recorded.[14]

- **The Kellington worm**, a Yorkshire specimen notorious for spitting venom at intruders in its marshy forest habitat. According to legend, a shepherd named Ormsroyd killed the snake but paid the ultimate price for his heroism. In its death throes, the serpent spat venom onto Ormsroyd's dog, which subsequently licked the shepherd's face and thereby killed him, even as the dog collapsed and died.[15]

- **The Lambton worm**, a Co. Durham-dweller ranked among Britain's most notorious dragons. According to legend, Sir John Lambton caught a small snake while fishing near his home and dropped it down a well. Lambton subsequently embarked on a Crusade to the Holy Land, returning to find that the snake had grown huge and voracious, establishing an island lair on the River Wear, from which it raided local farms for livestock. After constructing a suit of spiked armor, Sir John faced the reptile and killed it in a furious struggle.[16]

- **The Slingsby worm**, also from Yorkshire, whose legend closely mimics that of several other British reptiles. According to the standard tale, Sir William Wyvill - scion of an actual Slingsby clan in the 14th century - took his hound and a suit of armor spiked with razor blades to face the beast. He triumphed, but died like poor Ormsroyd in Sussex when the poisoned dog licked his face.[17]

- **The Sockburn worm**, another Co. Durham man-eater slain by Sir John Conyers sometime prior to the Norman Conquest of 1066. Like the hero of Bishop Auckland before him, Conyers presented his lucky sword to the Bishop of Durham as a keepsake. It was handed down to each new bishop in turn until 1826, although a facsimile replaced the original sword sometime in the 13th century.[18]

- **The giant adder of Alcester**, Warwickshire, was not a classic worm, but it proved deadly nonetheless for a victim who suffered its bite in 1544. Unfortunately, no further description of the reptile has survived.[19]

Wales has no shortage of dragon lore, but classic "worms" are few and far between. The Cardiff worm, said to drown swimmers in a whirlpool on the River Taff, may well have been illusory. A more intriguing tale comes from Snowdon, where an aquatic serpent supposedly attacked a group of 18th-century swimmers at Llyn y Gadair, devouring one of the men. Much more recently, in October 1988, a river-dwelling predator killed 35 sheep on the Bodalog farm, outside Rhayader. Although unseen, the Beast of Bodalog - which killed its prey by biting the animals' necks - was widely described as a serpent. Author George Eberhart nominates feral dogs or "an unknown species of giant mink" as more likely culprits.[20]

Scotland, meanwhile, produced more startling reports in autumn 1965. Witness Maureen Ford was riding with friends on the A85 near Perth, on 30 September, when she saw a long, gray serpentine creature crawling along the bank of the River Tay. Her published statement - "It had no ears, but I'm sure I saw long pointed ears" - is confusing at best. The next morning, motorist Robert Swankie saw the beast or its twin beside the same road, closer to Dundee. Swankie reports, "The head was more than two feet long. It seemed to have pointed ears. The body, which was about 20 feet long, was humped like a giant caterpillar. It was moving very slowly and made a noise like someone dragging a heavy weight through the grass." No reptiles known to science have protruding ears, yet Internet reports somehow managed to conclude that Ford and Swankie glimpsed the legendary *beither* of antiquity.[21]

Traffic in exotic reptiles explains most (or all) or Britain's modern giant-snake reports, as indicated by the following:

- 24 April 1992: Burglars returned a stolen python to its owner by shoving it through the letterbox of a library at Golborne, Lancashire.[22]
- 28 February 2000: A pregnant 14-foot boa, described in press reports as "one of only eight in this country," escaped from its owner in Wigston, Leicestershire.[23]
- 16 November 2000: A 19-foot python escaped from its owner's home in County Durham, but was soon recaptured by firefighters and RSPCA officers.[24]
- 26 August 2001: Members of a Woolwich musical group, the *New Wine Church Band*, found a six-foot python hiding in their van as they prepared to play at a London wedding.[25]
- 28 July 2002: Hikers found a 4-foot boa in woodlands near Woodfalls, Wiltshire.[26]
- 31 August 2002: RSPCA officers captured a 10-foot boa in a garden at Ilfracombe, Devon.[27]
- 25 September 2002: A couple walking their dog met a 15-foot "Burmese python or boa constrictor" in Nottinghamshire. Police and RSPCA officers failed to locate the reptile.[28]
- 21 January 2003: RSPCA personnel pulled a 14.5-foot python from the Rotten Park Road canal in Edgbaston, Staffordshire, afterward describing their catch as "like something out of a horror movie."[29]
- 26 July 2003: A motorist in Ashby, Leicestershire reported running over a black snake 6-8 feet long. He later returned to the scene but found no trace of the animal.[30]
- 17 February 2004: Bus passengers in Swindon found a boa constrictor aboard their vehicle, inside a plastic box. An address on the box led police to Park South, where a teenage resident admitted selling the snake to an unknown party for £25.[31]
- 15 March 2004: Thieves stole a 7-foot python from its owner's home in Dunfermline, Scotland. A month later, garage mechanics found the snake dead inside the dashboard of a car owned by Arnold Clark, of Halbeath. Police charged an unidentified suspect with theft on 15 April.[32]
- 10 August 2004: Police issued warnings after witnesses sighted a 6-foot gray snake behind a church in Slimbridge, Gloucestershire. The reptile eluded searchers.[3]
- 28 October 2004: Residents of Hove, East Sussex, posted signs announcing the escape of a pregnant Burmese python.[34]
- 28 February 2005: Authorities announced an animal-cruelty inquiry after two 7-foot boas were found dead in a trash bin at Snydale, near Wakefield, Yorkshire. Two dead rabbits were found with the snakes, sealed inside a plastic bag.[35]
- 1 August 2005: RSPCA officers found a 7-foot boa drowned in a river on Station Road, in Llanfairfechan, Wales. Results of an inquiry to determine whether the snake was deliberately killed remain unpublished.[36]
- 12 August 2005: Police appealed for help after a 5-foot boa escaped from its owner's home in Maryhill, Glasgow.[37]
- 4 September 2005: RSPCA officers captured a 12-foot Burmese python in woodlands near Oswaldtwistle, Accrington, Lancashire.[38]
- 17 October 2005: Police in Manchester abandoned their week-long search for a 10-foot boa constrictor that frightened residents in a block of flats. Neighbors claimed the snake was abandoned by its owner in August and lived in Manchester's sewers until cold weather drove it to seek warmer digs.[39]
- 9 April 2007: Authorities were baffled by the discovery of a dead python measuring 17 feet 8 inches on Newlands Lane in Cloughton, Scarborough. A police spokesman told BBC News, "We

don't know if it had been living there wild, or had just been dumped."[40]

- 16 April 2007: A 9-foot python, this one very much alive, was captured by police in woodlands outside Ledbury, Herefordshire. Officers dubbed the snake "Monty" but failed to locate its owner.[41]

- 13 June 2007: Dorset police caught a 5-foot python in the Bryanston district of Blandford, after a local woman saw it sunning on a woodpile.[42]

- 23 June 2007: RSPCA officers bagged a 12-foot reticulated python on a country road near Gwehelog, in Monmouthshire, Wales. Public appeals to the snake's presumed owner have thus far proved fruitless.[43]

Scandinavia

Snake species native to Scandinavia include the European adder, grass snake, and smooth snake. For reasons still unclear, adders appear to reach their record length in Sweden, at 35 inches, while smooth snakes peak around 32 inches. Grass snakes occasionally reach 47 inches, still well short of "giant" status.[44]

As in Britain, though, Scandinavian folklore suggests the existence of much larger serpents. The titanic *Midgardsorm* or *Jormungandr,* whose arched coils represented rainbows, was clearly a mythical beast, but the same may not be true for the Scandinavian "dragons" called *lindorms.*[45]

Descriptions of the *lindorm* vary widely and are thus somewhat confusing. Sweden's Olaus Magnus (1490-1557) and Erik Pontoppidan (1698-1764), bishop of Bergen, Norway, considered them sea serpents, though Pontoppidan wrote that they "are not generated in the sea, but *on land,* and when they are grown so big that they cannot move about on the rocks, they then go into the sea, and afterwards attain their full growth." Pontoppidan cited reports from Norse farmers of "great snakes" measuring "several fathoms" in length. Similar creatures also dwelled in various Scandinavian lakes - and are still reported today as lake monsters.[46]

More surprising yet are eyewitness reports of land-dwelling *lindorms* surviving into the late 19th century. Swedish folklorist Gunnar Olof Hylten-Cavallius published a book titled *On the Dragon, Also Called the Lindorm* in 1885, including 48 verbatim eyewitness reports collected since 1878. Summarizing those accounts, Hylten-Cavallius wrote:

> In Varend [southern Sweden] - and probably in other parts of Sweden as well - a species of giant snakes, called dragons or lindorms, continues to exist. Usually the lindorm is about 10 feet long but specimens of 18 or 20 feet have been observed. His body is as thick as a man's thigh; his colour is black with a yellow-flamed belly. Old specimens wear on their necks an integument of long hair or scales, frequently likened to a horse's mane. He has a flat, rounded or squared head, a divided tongue, and a mouth full of white, shining teeth. His eyes are large and saucer-shaped with a frightfully wild and sparkling stare. His tail is short and stubby and the general shape of the creature is heavy and unwieldy.[47]

According to Hylten-Cavallius, *lindorms* made their homes in caves and swamps, as well as lakes. Twelve of his reports involved specimens killed by villagers, but such attacks were risky. "When alarmed," he wrote, "[the *lindorm*] gives off a loud hissing sound and contracts his body until it lies in billows; then he raises himself on his tail four or six feet up and pounces on his prey." Hylten-Cavallius distributed posters offering cash for a *lindorm*'s remains, but got no takers.[48]

Nearly a century after Hylten-Cavallius published his treatise on *lindorms,* Swedish author Sven Rosen acknowledged, "There is no truly satisfactory explanation for these 19th-Century lindorm reports." He suggested "hallucinations such as those caused by epileptic fits," but was forced to discount that explanation in cases with multiple witnesses. "Many of the 31 additional cases with which I am familiar also had multiple witnesses," Rosen wrote. "One can speak of 'collective hallucination' without effectively explaining anything."[49]

No modern Scandinavian mystery serpents on land compare to the *lindorm* in size. Fortean authors Colin and Janet Bord cite two Danish cases, separated by 30 years, but neither snake qualifies as a giant. In September 1943 a 6-foot yellowish-brown water snake emerged from the River Gudenaa, in Jutland, to attack a passer-by without provocation, and put the man to flight after he failed to kill it with a stick. June 1973 brought reports of an even smaller specimen, some 4.5 feet long, that frightened witnesses on Mön Island, in the Ulushale Forest. The snake had dark dorsal scales, a pale belly, and a "nasty head." One witness thought it might have been "a giant eel taking a walk on land."[50]

Spain

Our reports from Spain are more recent than those from Scandinavia, but less impressive. Science recognizes 13 snake species in Spain, including two species - the Aesculapian snake and the Montpelier snake - which may exceed six feet in length. Neither, however, matches the description of the 6-to-7-foot green snake run over by a car near Chinchilla de Monte Aragón, Albicere Province, on 22nd July, 1969.[51]

Another 6-foot reptile, distinguished by its "huge head," made several appearances at a farm in Orihuela, Alicante Province, during June 1970, but it remains unidentified. The real Spanish shocker was a "monstrous serpent with a mane and a head like a baby's," sighted near Aceuche, in Cáceres Province, in July 1973. Sadly, the vague published descriptions leave us guessing as to whether its size was "monstrous," or merely its appearance.[52]

France

The only French dragon on record seems to be the Ballon Monster, a huge serpent thrown up by flood waters in Alsace, during 1304. Perhaps amphibious, the beast attacked humans and livestock alike, but no detailed descriptions remain and its fate is unknown.[53] Modern France harbors 10 native species of snake known to science, but the only giants on record are exotic imports, such as the 7-foot python which Nadege Brunacci found in her apartment's plumbing, in October 2007.[54] We must therefore turn elsewhere in search of true giants.

Italy

Pre-Christian poet Sextus Propertius (c. 50-45 BCE) described an Italian dragon which 19th-century author Charles Gould presumed to be a large python. As described by Propertius in his *Elegy VIII:*

> Lanuvium [25 miles from Rome, on the Appian Way] is, of old, protected by an aged dragon; here, where the occasion of an amusement so seldom occurring is not lost, where is the abrupt descent into a dark and hollowed cave; where is let down - maiden, beware of every such journey - the honorary tribute to the fasting snake, when he demands his yearly food, and hisses and twists deep down in

the earth. Maidens, let down for such a rite, grow pale, when their hand is un-
protectedly trusted in the snake's mouth. He snatches at the delicacies if offered
by a maid; the very baskets tremble in the virgin's hands; if they are chaste, they
return and fall on the necks of their parents, and the farmers cry, 'We shall have
a fruitful year.'"[55]

Pliny the Elder (AD 23-79), in the eighth book of his *Naturalis Historia,* claimed that large "boas" were
common throughout Italy, including one killed on Vatican Hill during the reign Emperor Claudius (AD
41-54) that was found with a whole child inside it. Irish author Ronan Coghlan suggests that the serpent
(s) of Lanuvium remained alive and well as late as 1651 A.D., but he provides no further details.[56]

Moving offshore, to the island of Sardinia, we find the legend of *scultone,* a "sort of basilisk" that terrorized villagers "for many a millennium." Again, details are sparse, but the stories may relate to much
more recent events from Sicily. In late December 1933 villagers from Syracuse spent two days tracking
an 11-foot snake which they called *colovia.* The hunters found their prey, killed it, and burned its carcass, fearing that it was an omen of disaster. Author George Eberhart suggests that the *colovia* was an
"escaped python or boa."[57]

More recently, in summer 1963, a 12-foot snake surprised residents of Udine, Italy with its size *and* its
ability to whistle. The reptile's length alone rules out all of Italy's 11 known snake species, and since it
was never captured, the creature remains unidentified.[58]

Greece

Modern pythons derive their name from *Pythō,* a huge serpent slain by the god Apollo, so it is no great
surprise that Greek dragons resembled the Scandinavian *lindorm.* Aristotle, in his *History of Animals*
(350 BCE) wrote that "the eagle and the dragon are enemies, for the eagle feeds on serpents." He also
observed that "the Glanis [a large freshwater creature, for which the wels catfish *Siluris glanis* is named]
in shallow water is often destroyed by the dragon serpent."[59]

Pliny relays a story from the Greek author Democritus (460-370 BCE) concerning Thoas, a resident of
Arcadia who raised a large serpent from infancy and then released it in the desert when it grew too large
to handle. His kindness was rewarded when robbers waylaid Thoas in the same vicinity and the reptile
came to save him, having recognized his voice. Although Pliny calls the beast a dragon, he goes on to
explain that "[t]he dragon is a serpent destitute of venom."[60]

Serbia

Moving north from Greece to Serbia, we find our first reports of truly giant snakes in Eastern Europe -
and some very recent ones, at that. In the 1930s, a shepherd named Obrad Vojvodic met a huge serpent
while searching for lost sheep, then ran home and lapsed into a state of shock. Shortly before the out-
break of World War II, witness Nenko Stanivukovic saw a snake exceeding 30 feet in length, near the
village of Zeljevo (located in present-day Republika Srpksa, Bosnia and Herzegovina).[61]

A half-century later, in the 1980s, Luka Stanic found a serpent's trail three feet wide, slashed through the
midst of his cornfield outside Jelasinova, in western Serbia. Researcher Aleksandar Lovcanski claims
that on 10th October, 1999 an unnamed Serbian newspaper reported another incident, involving a huge
snake seen swallowing a goat beside the Topcider River, on Belgrade's outskirts. Witness Mustafa Sen-
dic described a snake more than 30 feet long, coloured with black and greenish spots. Seven other wit-

nesses came forward with similar stories during October, including Djordje Milanovic, who saw the snake crawl across a road near the Topcider railway station. From descriptions of its colour, Lovcanski concluded that the reptile was a misplaced anaconda.[62]

The following year, in spring or summer 2000, author Marcus Scibanicus reports that a bus traveling past Ivanjica - 18 miles south of Ovčar Mountain, near Čačak - was forced to stop for a 33-foot snake crossing the road. Despite multiple witnesses aboard the vehicle, no clear description of the reptile is available.[63]

Hungary

Hungary's legendary *sárkánykígyó* was a giant winged reptile imbued with supernatural powers to control inclement weather, but its immature form - the *zomok* - was a much more down-to-earth creature. Ancient accounts describe it as a large swamp-dwelling snake that devoured pigs and sheep. The *zomok* possessed no special powers and was easily dispatched by peasant villagers with common weapons, suggesting actual encounters with large and as-yet-unidentified reptiles.[64]

Poland

Four snake species are recognized as Polish natives, including the Aesculapian snake that may exceed six feet in length, but none explain the serpent killed by a forester named Zander near Wroclaw, the capital of Lower Silesia, on 26th July 1713. In death, the reptile measured 17 feet 4 inches. It remains unidentified today.[65]

Ukraine

Ukraine harbors 25 indigenous reptile species, including several snakes, but none match descriptions of the huge serpents described in a newspaper article from 1841. According to that story, local Cossacks had encountered several massive reptiles dwelling in reed banks along the Dniester river, which flows from its source at Drohobych, near the Polish border, through Moldova and Ukraine to the Black Sea. Details are sadly lacking from the sole Ukrainian report, beyond a claim that the serpents killed man and beast alike, moving fast enough to catch riders on horseback.[66]

Turkey

The last leg of our European journey takes us to Turkey, historically standing astride the frontier between Europe and Asia. Eleven snake species are native to Turkey, including two that may exceed three feet in length, but others remain unaccounted for.[67]

Pliny, in his *Natural History,* tells us: "Metrodorus says that about the river Rhyndacus, in Pontus, [serpents] seize and swallow birds that are flying above them, however high and however rapid their flight." Unfortunately, at least 10 Greek authors shared the name Metrodorus between the 5th century BCE and the early 2nd century AD, nine of them known during Pliny's lifetime. Today, we can say only that the Rhyndacus River flows through Pontus, a Turkish region on the southern coast of the Black Sea.[68]

A more recent mystery surrounds the reptile killed by a forester from Dikmene, in Balikesir Province (western Turkey), in 1938.

A 16-year-old boy initially tried to kill the snake with a stick, whereupon it coiled around his body and began to squeeze. His cries for help brought reinforcements, who drove the snake off but failed to kill it. The hero who later accomplished that deed received a bite on one foot, but suffered no ill effects from it. The dead snake reportedly measured 11 feet 5 inches long and 16 inches in circumference. No further description is presently available.[69]

The Boas

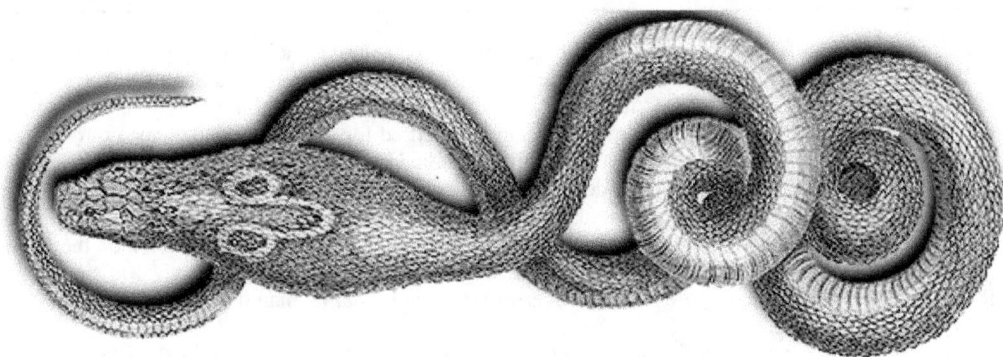

Chapter 3. Africa

From Europe, where snakes are relatively small and rare, we move on to Africa, whose 450 snake species comprise 16 percent of Earth's total.[1] Not only are they numerous, but Africa's crop includes one of the Big Six and two other pythons, besides.

Africa's certified giant is the rock python (*Python sebae*), which - with its subspecies *P. s. natalensis* - is found throughout the sub-Saharan continent. Confusion surrounds its original classification by Johann Gmelin (as *Coluber sebae*) in 1788, with the original type locality listed as "America." Alvin Loveridge claimed in 1936 that no type locality was given, but A.F. Stimson of the British Museum listed the type locality in 1969 as "Guiara [*sic*] Brazil." That muddled mystery aside, all sources agree that the rock python may exceed 20 feet. Clifford Pope cites a record length of 32 feet but offers no specifics, while Internet websites claim an "unconfirmed" record of 37 feet 9 inches. In the latter case, villagers reportedly devoured the python before authorities arrived to confirm its size.[2]

Pope certifies the rock python as a man-killer, citing two specific cases. Both occurred on Ukerewe, an island in Lake Victoria, where victims crushed by two different snakes, years apart, included a boy and a woman. In the latter case, the reptile measured 14 feet long and 16 inches in diameter. In neither case was the victim devoured, but a 10-year-old boy *was* swallowed by a rock python near Durban, South Africa, in November 2002. Another child witnessed the incident and ran for help, but the reptile eluded police who were slow to arrive on the scene. Craig Smith, from the Fitzsimmons Snake Park, examined the snake's tracks and judged its length to be between 18 and 20 feet.[3] Africa's other pythons, in order of size, are the ball or royal python (*P. regius*) and the Angolan python (*P. anchietae*). Neither attains "giant" size, with their record specimens falling around six feet in length. We may, therefore, exclude them from consideration as we survey reports of truly giant snakes from the Dark Continent.[4]

Classical Serpents

Our first report of giant snakes in Africa comes from Aristotle's *History of Animals,* begun around 343 BCE. In Book VIII he writes that "in Libya, the serpents, as it has been already remarked, are very large. For some persons say that as they sailed along the coast, they saw the bones of many oxen, and that it was evident to them that they had been devoured by serpents. And, as the ships passed on, the serpents attacked the triremes, and some of them threw themselves upon one of the triremes and overturned it."[5]

We have more details for the next report, logged from Egypt during the reign of Ptolemy II (281-246 BCE). At some point during those four decades, a 45-foot serpent was captured alive and carried to Alexandria, where its keepers "tamed" it by starving it into submission - a technique *not* recommended for potential python-handlers. While no point of origin is specified for the snake, it likely came from Ethiopia, where Greek historian/geographer Agatharchides of Cnidos claimed that serpents grew large enough to battle elephants.[6] Our next and most detailed account comes from 256 BCE, in the eighth year of the Punic Wars, when Roman general Marcus Atilius Regulus led an expedition against Carthage. After crushing the Carthaginian fleet at Cape Ecnomus, Regulus staged an amphibious invasion on the coast of modern-day Tunisia and thrust inland, sweeping all before him.[7] However, on arrival at the Bagrada River (now the Medjerda), Regulus met an adversary more intimidating than the Carthaginian army.

The enemy in question was a titanic serpent which rose from undergrowth along the river's opposite bank to confront the Roman legionnaires. It made no effort to attack until the first troops crossed the river, whereupon it swiftly killed four men in turn. Regulus then ordered up his siege ballistae, resembling huge crossbows, and bombarded the reptile with boulders until its skull was crushed and any sign of life had been extinguished. Soldiers skinned the snake and sent its 120-foot hide back to Rome, whereupon Regulus was honored with an ovation (normally reserved for victory attained in civil war). Even allowing for the standard 30-percent stretch factor, the trophy still represents a snake exceeding 80 feet. The skin remained on display in a temple on Capitol Hill until 133 BCE, when it was somehow lost during Rome's Numantine War with the Iberian Celts. Since that war was fought entirely on Spanish soil, the loss of such a trophy remains unexplained. Reports that Pliny saw the hide are clearly incorrect, since its loss occurred 156 years before his birth.[8]

Other reports of giant snakes from classical antiquity include the writings of first-century Greek geographer Artemidorus Ephesius, who - according to his colleague Strabo (64 BCE-24 AD) - "mentions serpents of 30 cubits [45 feet] in length, which can master elephants and bulls. In this he does not exaggerate; but the Indian and African serpents are of a more fabulous size, and are said to have grass growing on their backs." Pliny described Ethiopian "dragons" 30 feet long, while Roman historian Suetonius (71-135 AD) reports that a snake displayed in Rome during the reign of Augustus (27 BCE-14 AD) measured 75 feet.[9]

German explorer Job Ludolphus, in his *New History of Ethiopia* (1684) declares that the nation's "dragons are of the largest size, very voracious, but not venomous" - which suggests giant pythons. A later document, Volume 1 of *Harris's Voyages* (1744), quotes explorer John Leo's observations from the Atlas Mountains of northwest Africa, spanning the modern nations of Morocco, Algeria and Tunisia. There, dwelling in caves, Leo found "many monstrous dragons which are thick about the middle, but have slender necks and tails, so that their motion is but slow." Leo's account might well describe a python that had recently consumed large prey.[10]

An earlier report from the Atlas Mountains comes from Joannes Leo Africanus (né Hasan ibn Muhammed al-Wazzan al-Fasi), born at Granada, Spain, in 1488. Captured by Mediterranean pirates as a young man and sold into slavery, then liberated by Pope Leo X (1513-21), Africanus traveled widely through northern Africa before his death at Tunis, in 1554. His reports of huge serpents prowling the Atlas range predates John Leo's, but Africanus maintained that the reptiles were venomous.[11]

Modern Monsters

Reports filed from Algeria during the latter half of the 20th century suggest not only that early reports of the region's huge serpents were true, but that some of the reptiles still survive - or did, as late as 1967. The first modern report dates from 1958, when a native Algerian serving with French occupation troops at Beni Ounif, one Belkhouriss Abd el-Khader, was attacked and bitten by a 43-foot snake. His fellow soldiers killed the reptile and preserved its skin, but it was later lost.[12]

A year later, at Aïn Sefra, nomads trapped a 120-foot snake that proved sluggish after devouring one of their camels. A runner summoned soldiers of the French 26th Dragoons, commanded by Captains Grassin and Laveau, who found the snake penned in a trench topped with branches. Rifle fire only annoyed the huge beast, so machine guns were brought up to finish the job. A distinctive scaly crest marked the dead reptile's head. No published story explains what was done with its carcass.[13]

Five years after French troops finally departed from Algeria, in early January 1967, a construction worker named Hamza Rahmani met a smaller crested snake, this one some 30 feet long, while working on the Djorf-Torba dam, east of Béchar. Rahmani killed the snake with his bulldozer blade, later reporting that its teeth measured 2.4 inches long. Near the end of 1967 Rahmani saw a second large snake at the same building site but failed to kill that one, estimating its length at 18-23 feet.[14]

The *Lau*

Some confusion surrounds the giant serpents of Sudan - namely, concerning whether they are snakes at all. The same debate exists concerning other large reptiles of the Dark Continent, and so we may as well address them in relation to the *lau*. British troops occupied Egypt and Sudan in 1882, remaining in the latter country until 1956. In 1923 Deputy Governor H.C. Jackson published a report on the Nuer tribal confederation of Southern Sudan and western Ethiopia, which comprises one of the largest ethnic groups in East Africa. That report included a description of the *lau*, a giant snake said by the Nuer to inhabit swamps around the source of the White Nile. According to Jackson, "Various fantastic attributes attach to it, which may be mere embellishment on the part of the Nuer without necessarily disproving the existence of some hitherto unknown serpent of exceptional size."[15]

> Thus [Jackson wrote] some say it has a short crest of hair on the back of its head not unlike that of a crowned crane: others that it has long hairs, reminiscent of some of the mud-fish of the Nile, with which it entangles its unwilling victims and drags them into the river. In certain years, particularly the rainy season, its belly is said to gurgle like the rumbling of an elephant. The noise was heard in the year 1918 in the Bahr-el-Arab.
>
> The Nuer state that it inhabits holes in the banks of rivers or swamps, and spends most of its time in swamps. They are in mortal dread of it, and if they see the furrow in the ground that announces its presence, they run as fast as they can in the opposite direction. If a Nuer sees this serpent before it sees him, all is well, but if the serpent happens to sight even a large party of them first, all are expected to die.
>
> The Addar swamps extend for some 1800 square miles: those of the Bhar-el-Ghazal and its tributaries must cover an area of many thousands of square miles. Neither of these vast morasses has ever yet been explored, and many parts of them have never even been visited. It is therefore not improbable that some addition to our zoological knowledge may ultimately come from these places.
>
> There are so many stories of this creature from places as far apart as Bahr-el-Arab, Addar swamps, Bahr-el-Ghazal and Bahr-el-Zeraf, that it is difficult to dismiss as untrue the existence of some hitherto unknown serpent not unlike a gigantic python. The existence

of a bone or hornlike substance near the tail would illustrate a step further back in the development of the serpent creation, the small rudimentary legs of the python marking an intermediate stage between the Lau and the ordinary snake as we know it. The common factor in the various accounts given of this creature is that it differs in colour from a python, in the shape of its jaws, in its great length and girth, and there is no doubt that it is held in the greatest dread by the Nuer. They have a definite name for it, and there seems no particular reason, if this serpent were not distinct from the python, why the Nuer should fear a 12-foot Lau more than a 20-foot python.[16]

We recognize that neither birds nor fish have hair, but Jackson's report fired the imagination of British naturalist J.G. Millais, who was then engaged in tracing the source of the Nile. Millais in turn located one Sergeant Stephens, whose service with the occupation forces in Sudan had persuaded him of the *lau*'s existence. Millais found Stephens in Malakal, now capital of the Sudanese *wilayat* (state) of Upper Nile, and obtained the following statement:

> The Lau is said to inhabit the great swamps of the Nile valleys from below Malakal to Rejaf and Lake No to Shambe. The natives, whether Shilluk, Dinka or Nuer, have the same name for the great serpent. It is said to be extremely rare and seldom seen by man more than once a lifetime. Natives who are said to have seen a dead one describe it as 40 to 100 feet long, with a body as big as a donkey or a horse. The colour is said to be brown or dark yellow, and not green and black like the python. On its vicious-looking, snake-like head it has large tentacles, or thick, wiry hairs, with which it reaches out and seizes its victims. If a man sees a Lau first, the creature at once expires, but if the Lau is the first observer the human dies on the spot.
>
> In the time of Bimbashi B (a well-known telegraph inspector), one Abrahim Mohamed, in the employ of the company, saw a Lau killed near Raub at a village called Bogga. The man I knew, and closely questioned. He always repeated the same description of the monstrous reptile. More recently one was killed by some Shilluks at Koro-a-ta beyond the Jebel-Zeraf (Addar Swamps). I obtained some of the neck bones of this example from a Shilluk who was wearing them as a charm. These I sent to Deputy-Governor Jackson (now of Dongola province), who in turn sent them to the British Museum for identification, but no satisfactory explanation was given, nor was it suggested what species of snake they might belong to.
>
> Abrahim's story of the size and shape of the great reptile was corroborated by one Rabah Ringbi, a Niam-Niam from the neighbourhood of Wau in the Bahr-el-Ghazal, who had also seen a similar monster killed in some swamps at that place.
> Dinkas living at Kilo (a telegraph station on the Zeraf) told me that the Lau frequents the great swamp in the neighbourhood of that station and they occasionally hear its loud booming cry at night.
>
> A short time ago I met a Belgian Administrator at Rejaf. He had just come up from the Congo, and said he was convinced of the existence of the Lau, as he had seen one of the great serpents in a swamp and fired at it several times, but his bullets had no effect. He also stated that the monster made a huge trail in the swamp as it passed into deeper water.[17]

Millais noted that local tribesmen knew the common python as *nyāl*. He also questioned H.C. Jackson concerning the *lau* vertebrae, whereupon Jackson told Millais they were not unusually large and might belong to a normal python.[18] The facial tentacles are curious, and might be pure invention, but we shall see them again on another huge snake - in the American Midwest, of all places.

Captain William Hichens, Native Magistrate at Lindi, in what is now Tanzania, procured a mask carved in the likeness of a *lau*'s head by Mshengu she Gunda of Iramba (one of four districts in Tanzania's Singida Region) and published a photo of it in *Discovery* (December 1937). Later, Hichens wrote to author Frank Lane: "Mshengu lived on the south-east side of the Wembare swamps and, as a young man, had travelled and hunted around and over Victoria Nyanza and the Nile swamps. To my suggestion that there was no *lau* he said, 'I might have said, as a young man, when I was ignorant, that there is no such thing as a motor car. I had never seen or even heard of one then. But there is your motor car in the sight of my eyes and I have sat in its chairs and heard its bowels digest inside it. It is thus of the *lau*."[19]

Bernard Heuvelmans and others argue that the *lau* is not a snake at all, but rather a relict dinosaur.[20] That supposition seemingly depends entirely on the reference to its "loud booming cry" - which, in fact, is not linked to an eyewitness sighting. Conversely, not a single witness mentions any glimpse of legs or ever classifies the *lau* as anything but a gigantic snake.

A Giant Cobra

We have already seen reports of large crested snakes in Algeria and Sudan, but our next cryptid carries that trait to extremes - and seemingly inhabits a much wider range. William Hichens addressed it in his 1937 *Discovery* article, writing: "A beast over which controversy rages at this moment is the 'crowing crested cobra,' which, the natives say, is a snake, like a cobra, with a crest on its head and a loud, distinct cry like the crow of a cock."[21]

Science recognizes 15 species and subspecies of cobra in Africa. The largest are the banded water cobra (*Boulengerina annulata*) and Ashe's spitting cobra (*Naja ashei*), which may reach nine feet in length, but none rivals the crowing crested cobra's alleged length of 20 feet. In fact, the world's largest cobra and largest known venomous snake, the Asian king cobra (*Ophiophagus hannah*), holds a record of 18 feet 4 inches. No cobra - or any other snake, for that matter - displays a rooster-like crest or produces the sounds attributed to crowing crested cobras.[22]

This mystery serpent inhabits a broad range, from Tanzania southward through Zambia, Malawi, Mozambique to KwaZulu-Natal in South Africa. It is known by many names throughout its range, including *bubu,* on Mozambique's Lower Zambezi River; *inkhomi* ("the killer"), *kovoko, ngoshe* and *songo* ("strikes down on head") by various Malawi tribes; and *mbobo* in Zimbabwe. Animal collector Charles Cordier compounded the mystery in April 1973, with his report of large red-crested snake said to inhabit the Congo River Basin. Bernard Heuvelmans suggests that the same snake may be present in the Central African Republic, known to Baya tribesmen as *n'gok-wiki*. More recently, in 2006, members of an expedition mounted by the Centre for Fortean Zoology discovered that a Gambian reptile called *ninki-nanka,* described in prior accounts as a "dragon," was actually known to native tribesmen as a huge crested snake.[23]

Despite the variance in local names, descriptions of the serpent are remarkably consistent. Witnesses invariably note its great size, the scarlet crest atop its head with face to match, made doubly striking by its overall buff-brown to gray-black colour, and a fondness for climbing trees. It seems improbable that villagers would handle such a large and lethal snake, much less subject it to the probing examination required to determine sex in snakes, yet native tales maintain that only male crested cobra's sport distinctive turkey-like wattles and crow like roosters, while females produce a clucking noise reminiscent of hens. And if this is not strange enough, *both* sexes reportedly utter warning cries to potential enemies, described by those who have heard it as "chu-chu-chu."[24] In short, the snakes resemble no species of cobra known to science. They have even sacrificed the trademark cobra hood, it seems, in preference to crests and wattles. Legend has it that the crowing crested cobra dines exclusively on maggots, killing larger animals simply to let their flesh rot and produce the insects that it craves, but as Dr. Karl Shuker observes, such a diet would hardly sustain any 20-foot reptile. Bernard Heuvelmans relayed a more likely observation in 1978, from veterinarian Dennis Walker in Southern Rhodesia (now Zimbabwe). Dr. Walker claimed that the crested cobras found in his bailiwick preferred to eat hyraxes, mammals distantly related to elephants that commonly weigh between 5 and 10 pounds.[25]

A smaller version of the crowing crested cobra may exist in South Africa, according to game-ranger Harry Wolhuter's 1948 memoirs. Wolhuter described a 12-foot venomous snake called *muhlambela,* which sports a crest of "feathers" and bleats in imitation of a deer before striking at its victims from the limbs of trees. George Eberhart suggests that Wolhuter, despite his presumed familiarity with South African wildlife, mistook a molting black mamba (*Dendroaspis polylepis*) for a new and nonexistent species.[26] What could the crowing crested cobra be? Ex-game warden Charles Pitman, in his *Report on a Faunal Survey of Northern Rhodesia* (1934), considered it a fantasm concocted from garbled accounts of other venomous serpents, including the black mamba and brightly-coloured Gaboon viper (*Bitis gabonica*). George Eberhart adds two more snakes - the puff adder *(Bitis arietans)* and rhinoceros viper (*Bitis nasicornis*) - to the mix, although neither exceeds six feet in length.[27] A specimen would surely solve the riddle, and Dr. J.O. Shircore - Tanganyika's Director of Medical and Sanitary Services during World War II - professed to own a partial skeleton. Writing to the Oxford journal *African Affairs* in 1944, Dr. Shircore said:

> Its skeleton consists of a thin laneolate plate of bone (1.5 ins long by 0.5 in wide at its broadest part) with a markedly rounded smooth ridge, 0.5 in wide, slightly overhanging both sides of the upper border, with a distinct voluted curve to the left. The lower border is sharp-edged and faintly ridged. The lateral surfaces are concave, throughout the long diameter. The whole fragment is eminently constructed for the insertion and attachment of muscles - much the same as the structure of the breast-bone of a bird. Some skin, part of which, spread smoothly above the base of the plate, on one side, is red in colour: and attached to the lower angle is a dark wrinkled bit, which appears to be a remnant of the head-skin - all of which should be valuable for purposes of identification. A small portion of the bone, tapering towards both ends, 0.5 in long by 0.5 in wide, is missing from the lower anterior border, including the tip - it was broken off for use as medicine by the witch-doctor, from whom the specimens were obtained.[28]

A second note from Dr. Shircore indicates that he had obtained two other specimens. One included five lumbar vertebrae measuring 7mm long and 5mm wide, each with a concave articulating face 2mm high and 3mm wide; two ribs measuring 26mm along their curve; a patch of skin measuring 3 square inches; the skin-tip of the reptile's crest, 6mm long and 3mm wide at its base. The second specimen, a single vertebra allegedly belonging to a large crested cobra which had killed a man, measured 8mm by 9mm.[29]

Another crowing crested cobra may have been killed in May 1959, near Kariba in Zimbabwe's Mashonaland West Province. John Knott ran over a black 7-foot snake with his car and stopped to observe the reptile before it died, noting that its head bore a symmetrical crest supported by five bony spines. Karl Shuker notes that similar crested snakes, though much smaller in size, were reported from several Caribbean islands during the 19th century. Chinese reports also describe "rooster-crest snakes", but none rival Africa's mystery serpent for size - and none have been identified.[30]

Nguma-monene

Another reptilian cryptid mired in confusing reports is the *nguma-monene,* reported from eastern Cameroon and the neighboring Republic of Congo. Its name means "large boa" in the Lingala language, and while no witness to date has described the beast as possessing legs, some researchers persist in calling it a relict dinosaur.[31] Our first account of the *nguma-monene* comes from the Republic of Congo's northern frontier in 1961. The eldest sister of Michel Zabatou, then first secretary of the nation's General Assembly, was swimming in the Motaba River - a tributary of the Ubangi - when a snakelike head and neck broke the surface 50 feet away. Her cries brought other villagers to the river's bank, and all present clearly saw the reptile's forked tongue flicking rapidly as it swam upstream. The witnesses also described a frill or ridge on the animal's back.[32] Ten years later, Rev. Joseph Ellis saw a similar creature near the same site. While traveling by boat on the Motaba, to reach a village where he was supposed to teach a Bible class, Ellis met a huge reptile swimming across the river. He estimated its visible length at 30 feet, and noted a ridge of diamond-shaped serrations running the length of its spine. Residents of the nearby village shrugged off his report and refused to discuss it, leaving Ellis with an impression that the subject was "somewhat taboo."[33]

Dr. Roy Mackal notes that while forked tongues are displayed by snakes and certain lizards, no serpent known to science has a bony ridge along its back. He suggests that the *nguma-monene* may be a relict *Dolichosaurus,* a marine reptile from the Upper Cretaceous Period regarded by some paleontologists as an intermediate species between lizards and snakes. As an alternative suspect, Mackal proposes an unknown species of giant monitor lizard.[34]

The *Pumina*

No one suggests that the *pumina* is a lizard. In fact, one of its alternate names in the Democratic Republic of the Congo is *moma* ("python").[35] And while large rock pythons are common in the DRC, science refuses to accept that any reach the massive size attributed to the *pumina.* British seaman John M'Leod, in his *Narrative of a Voyage in His Majesty's Late Ship Alceste* (1817), describes his period of several months' captivity at Whydah on the West African coast (now Benin). While there, he observed pythons "double the length" of 36-footers reported from the Dutch East Indies, but no further details are offered.[36] The largest specimen on record allegedly surpassed the monster killed by Regulus and his legionnaires during the Punic War. According to an Internet website accessed in 2004 and no longer in existence, Congolese tribesmen killed a 130-foot python in 1932, then cooked and devoured it before Belgian colonial authorities could examine the titanic carcass.[37] The next *pumina* surfaced in 1959, and it seemed almost puny beside the monster reported 27 years earlier. Colonel Remy Van Lierde, a Belgian army officer, was patrolling the Congo's strife-torn Katanga district by helicopter when he saw a large snake crawling over open ground below. He estimated its length at 40-50 feet, with a head measuring three feet long and two feet wide. As Van Lierde dipped lower to examine the reptile, that fearsome head rose 10 feet off the ground and prompted him to reconsider. Nonetheless, Van Lierde (or a mechanic named Kindt, in some accounts) snapped a photo of the snake which has been widely published over the ensuing decades.[38]

Internet author Dave Juliano, writing on his Shadowlands website, writes that: "Experts have analyzed the pictures and have verified them as authentic. They also have verified the size of the creature by matching ground features to the snake."[39] In fact, though, only one photo has been published and nothing resembling an expert analysis has yet come to light. No definite location for the sighting is available, and the "ground features" - while clearly visible - provide no sense of scale. An anonymous critic, writing to the *Unexplained Mysteries* forum on 11th April, 2006, declared:

> Question has always been, how can you tell how big is the snake or worm in this very old picture? I have the answer to the old question, on the bottom, a little right of the picture is a mushroom. Seems like no one ever noticed it and even the Belgian pilot didn't when he took the picture. Unless anyone who knows [*sic*] of a giant mushroom, the creature on the picture is not quite as big as claimed.[40] In fairness to Van Lierde, wherever he may be, the "mushroom" is by no means clearly visible and may be a product of the viewer's imagination. **

The year after Van Lierde's sighting, a 60-foot python was seen in the Sangha River. William Gibbons identifies the witness as Pascal Mokengui, "employed by the former French Colonial Government to keep track of the civilian population in the remote villages situated in the swamps." History tells us that French Equatorial Africa - including the territory of Gabon, French Cameroun, the Republic of Congo, the Central African Republic, and Chad - formally dissolved in September 1958 to become the Union of Central African Republics, then gained complete independence as separate nations in August 1960. The Sangha River is a tributary of the Congo, flowing through Cameroon, the Republic of Congo, and the Central African Republic. Beyond that, we have no means of dating or locating Mokengui's sighting.[41]

South Africa

Aside from child-eating pythons and crowing crested cobras, South Africa claims two other great serpents. One, known as *blessie* ("bald one"), is described as brown, 15-50 feet long, and fond of hiding in an unspecified "deep pool beneath a tiny waterfall." Legend has it that the snake guards a sunken cache of diamonds, which no one ever sees, and it is fond of snatching children or cattle who linger at the water's edge.[42]

South Africa's last giant snake is a one-off, reported only in April 2000. It briefly terrorized villagers at Ezitapile, in the Lukholweni administrative area in Maluti, Eastern Cape Province. Witnesses described it as yellow, with a head resembling a horse's and a mane to match, a body as thick as a 5-gallon drum, and a long tail. No estimates of overall length were published, and George Eberhart may be correct in suggesting that the serpent was a rock python caught midway through shedding its skin.[43] By the same token, Eastern Cape Province is known for its wildlife and relatively unspoiled terrain. Tourist advertisements for the region call its "Wild Coast" an "enchanted coastline, too remote for modern development, that has remained virtually unspoilt and offers secluded bays and beaches." It is "one of South Africa's most unspoilt areas, a vast stretch of undulating hills, lush forest and spectacular beaches." In fact, local promoters say, it is nothing short of "a land that has been lost in time."[44]

What waits there to surprise unwary humans?

Who can say?

** EDITOR'S NOTE: The 'mushroom' mentioned here is probably a mushroom shaped termite mound. These are common in central Africa and can grow quite large. If this is the case it would indicate that the snake in the picture is indeed quite big. Remy Van Lierde died on 8th June 1990 after a distinguished career.

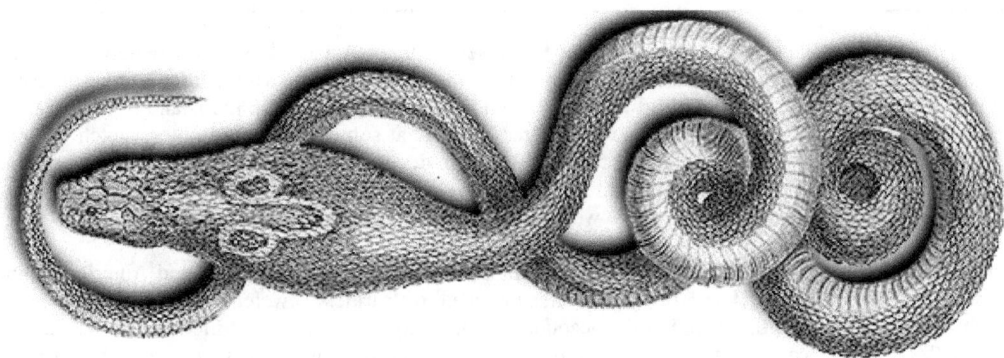

Chapter 4: Asia

Asia is the largest continent on Earth. Sprawling from the Ural Mountains and Suez Canal eastward to the Pacific Ocean and Philippine Sea, southward to the Indian Ocean, it includes 29.4 percent of Earth's total land area and contains more than 60 percent of the planet's total human population. Its geography is diverse, including the world's tallest mountains (the Himalayas), a cold desert (the Gobi) sprawling over 500,000 square miles, and some of Earth's oldest rainforests (covering 1,112,000 square miles), plus more than 20,000 named islands.[1]

We should not be surprised, then, to hear numerous reports of giant snakes emerge from Asia. Some apparently refer to known species, while others certainly do not.

Known Giants

Asia's greatest known giant is the regal or reticulated python, generally considered the longest - if not the heaviest - serpent on Earth. It inhabits Southeast Asia from the Nicobar Islands, Burma, Thailand, Laos, Cambodia, Vietnam and Malaysia, through Indonesia, the Indo-Australian Archipelago, and the Philippines. Clifford Pope credits it with an "accepted size" of 33 feet, and while a specimen 32 feet 9 inches long was reportedly killed on Sulawesi in 1912, no proof exists.[2]

Cryptozoologists shared cautious optimism in December 2003, when the Indonesian newspapers *Republika* and *Suara Merdeka* reported that a 49-foot regal python weighing 992 pounds had been caught in Sumatra's Jambi Province and placed on display at a zoo in Curugsewu, on Java. The snake - named *Kembang Wangi* ("Fragrant Flower") by its keepers - reportedly devoured four dogs per month. Early reports also claimed that Fragrant Flower had been examined by experts from Gajah Mada University, the Institute of Sciences of Indonesia, and the state Institute of Agriculture at Bogor. If so, they were somehow deceived. When finally measured in March 2004, the snake was "only" 22 feet long and tipped the scales around 250 pounds.[3]

Despite the reticulated python's great size, allegations of man-eating are rare and poorly documented. Clifford Pope noted that "virtually no evidence for the killing or eating of human beings by a giant snake is of the indisputable, eyewitness type," but he presented three Asian cases "most widely credited by herpetologists." Austrian naturalist Felix Kopstein (1893-1939) reported two fatalities from Indonesia in the early 20th century: a 14-year-old boy swallowed by a 17-foot python on the island of Salibabu, and an adult woman eaten by "a still larger reticulated python" at some unspecified place. Dutch naturalist Louis Coomans de Ruiter also claimed two women were devoured "in the same part of the world," but Pope's brief summary of that case fails to indicate if they were killed at the same time or in separate incidents.[4] A python that could swallow two women in one sitting must truly be a giant.

Two other cases were reported by Austrian zoologist Franz Werner (1867-1939) and American animal collector Frank Buck (1884-1950). Werner cites the death of Burmese jeweler Maung Chit Chine, allegedly devoured on a hunting trip, but the account cannot decide if Maung was killed before World War I or in 1927. Buck's case, from 1932, involved a Filipino youth swallowed by his father's 25-foot pet reticulated python. More recent cases include an 8-year-old Burmese child allegedly eaten by a 20-foot python in 1972; a 29-year-old Malaysian rubber tapper, found dead in September 1995 with his head stuck in a python's mouth; and a 38-year-old woman killed by a "10-foot python" which was, in turn, killed while trying to eat her in November 2003.[5]

The Internet provides photographic documentation of three alleged man-eating pythons. One offering apparently depicts the 1995 Malaysian victim, while the others - presented as alleged photo sequences - are suspect at best. Both offerings show different pythons engorged with their latest meals, then depict their opened bodies containing human remains.

Contradictory captions attached to the photos on various websites place the incidents in Borneo, Singapore and South America, identifying one victim as a local child, a camper, or "a crew member on an oil rig." The other victim, a bearded male, remains unidentified. Critics accept the early shots as genuine depictions of captive pythons, while suggesting that the final frames are fake.[6]

If in fact the reticulated python is Earth's largest snake, it has competition for that honor in its own backyard. The Indian python (*Python molurus*) and its Burmese subspecies (*P. m. bivittatus*) jointly hold title to a slot on the Big Six, although their maximum size is disputed.

Clifford Pope grants a maximum length of 20 feet, while adding that "no valid evidence of specimens more than 20 feet long has appeared." *National Geographic,* meanwhile, credits Burmese specimens with a record length of 23 feet, and a Burmese python at Serpent Safari Park in Gurnee, Illinois reportedly measured 27 feet in 2005. *P. molurus* inhabits a broad range including Pakistan, India, Sri Lanka, southern Nepal, Bangladesh, Myanmar, southern China, Thailand, Laos, Vietnam, Cambodia, Peninsula Malaysia and various Indonesian islands.[7]

Asia claims four other pythons species, including *Python curtus* in Southeast Asia, *P. c. breitensteini* on Borneo, *P. c. brongersmai* on the Malay peninsula, and *P. timoriensis* on the Lesser Sunda Islands. None are giants, and *P. c. breitensteini* is the largest of the lot, with a record length of seven feet.[8]

One other Asian giant falls outside the family of pythons. The king cobra (*Ophiophagus hannah*) is the world's longest known venomous snake, credited with a record length of 18-18.5 feet in various reports. Its documented range spans the territory from northern India to southeastern China, the Malay peninsula, Thailand, western Indonesia and the Philippines.[9]

At this stage in our global quest, it comes as no surprise to learn that giant snakes outstripping any known species in size have been reported throughout Asia, or that some appear in regions where large pythons and king cobras are officially unknown. We shall begin our Asian tour in the frozen north and travel on from there in search of monsters still unrecognized.

Siberia

Most readers in the West know Siberia only from its Cold War reputation as a frigid dumping ground for communism's political prisoners. Most do not know that Siberia spans most of northern Asia and the modern Russian Federation, stretching eastward from the Ural Mountains to the Pacific Ocean, and southward from the Arctic Ocean to the borders of China, Mongolia, and Kazakhstan. It covers 5,056,600 square miles - some 77 percent of Russia - yet contains only 30 percent of Russia's population.[10]

While it is true that Siberia suffers harsh winters - summer in the far north only lasts a month, and Omyakon (in the Sakha Republic) qualifies as Earth's coldest town, with a record temperature of -96.1° F - summer temperatures in southern Siberia range from 97-100° F. Curiously, Siberia's reports of giant snakes emanate from the Primorsky Krai district, bordered by China, North Korea, and the Sea of Japan, where average annual temperatures range from 34-41° F - hardly ideal for active reptiles.[11]

Fortean author Paul Stonehill - a Ukrainian native who emigrated to the United States at age 13, in 1972 - reports "numerous" sightings of huge snakes around Lake Tonee and the Samarga River, in Primorsky Krai. Witnesses describe serpents ranging from 15-30 feet in length, far surpassing the size of any known Russian reptile. In 1984, a group of workers en route to their jobs at the Nikolayevskiy pyrrhotite mine, at Dal'Negorsk, allegedly saw a 30-footer cross the road in front of them. Russian researcher Alexei Sitnikov led a team to Lake Tonee in 1993 but found no snakes - although they did report encounters with a hairy "snow man."[12]

Mongolia

This nation of 603,749 square miles borders Russia to the north and China to the south. It is the least densely-populated independent country on Earth, averaging 4.8 persons per square mile. Even that statistic is misleading, though, since 38 percent of Mongolia's 2.9 million citizens live in Ulan Bator, the capital city. Nationwide, 30 percent of all Mongolians are rootless nomads. The brutal Gobi Desert covers much of southern Mongolia, boasting a range of temperatures from -40° F in winter to 122° F in summer.[13]

In May 2005 CFZ Zoological Director Richard Freeman led an expedition to the Gobi, seeking Mongolia's elusive "death worm" and a 6-foot "horned snake" unknown to science, and returned with eyewitness reports of huge "dragons." One resident of Ulan Bator, employed as a mining firm's computer programmer, related a second-hand report from the mid-1940s.

According to this man, his aunt lived in a rural village near Mongolia's northern frontier at the end of World War II, and she had personally seen the carcass of a huge serpentine creature frozen in a nearby river. Only the upper portion of the "dragon's" scaly body was exposed, measuring 98 feet. The hungry villagers devoured most of it before spring thaws and flooding washed the rest away.[14]

Upon reaching the Gobi proper, Freeman's team heard more dragon tales, including that of a physician based in Ulan Bator, who met one of the monsters while visiting the desert town of Bulgan Sum in 2004. This reptile was no hundred-footer, but it was alive and well, looped in green coils at the bottom of a village well. The doctor described it to Freeman as a "Chinese dragon," with visible limbs.[15]

China

In fact, descriptions of dragons in China were seldom consistent. The reptiles were said to evolve through 3,000-year life-cycles, spending their first 500 years as small water snakes, then blossoming into huge serpents resembling Scandinavia's *lindorms,* sprouting legs 1,000 years later, growing horns around age 2000, and finally developing wings at the end of another millennium. The second phase, in fact, may have represented actual snakes of some yet-unidentified species.[16] One such reptile, the giant Pa snake, is depicted in the *Shan Hai King,* a classic volume of Chinese mythology written around 2250 BCE. According to that volume, Pa snakes were large enough to swallow elephants. They infested the Ta Hien Mountains and rendered the range uninhabitable for humans. Other giant snakes, belonging to a different species, roamed at will through the Siong Jan Mountains. Smaller serpents, but none the less deadly, were also reported from Quansi Province (now southern China's Guangxi Zhuang Autonomous Region). Writing in 1670, German Jesuit scholar Athanasius Kircher (1602-80) described snakes ranging from 10 to 30 feet long, the larger specimens known for dining on children. The largest serpent recognized from modern China is *Python molurus,* ranging from Sichuan and Yunnan Provinces eastward to Fujian, Hainan, and Hong Kong.[17]

Nepal

Author William Gibbons reports a unique "dragon" tale from Nepal, which in fact seems to describe a huge - and incredibly sluggish - serpent. The event allegedly occurred in 1980, was relayed to Gibbons in 1998, and was not published until 2003, nearly a quarter-century after the fact. The story comes from Ward Satterlee, a Minnesota preacher who, in turn, heard the tale from Indian missionary Resham Poudal, while Satterlee was traveling in Nepal.[18] As reported third-hand by Gibbons, Poudal and several companions camped out in a jungle valley one night, in a clearing beside a huge log. Early next morning, as they prepared to break camp, the "log" roused itself and began creeping away at a snail's pace.[19] Poudal told Satterlee:

> The animal was at least 40 feet in length, and about 5-7 feet in circumference. It was extremely well camouflaged and was a greenish brown in colour. Large armored scales protected the underbelly, and my informants tell me that the "dragon" moves very slowly - barely a foot or two per day.

> Its jaws were huge, and we were warned not to step directly in front of the creature, as it was capable of performing a powerful inhalation - sucking its prey (water buffalo) into its cavernous jaws from a distance of 10 to 15 feet. A six feet tall man could easily stand upright within its open jaws, which were enormous. Its eyes take on a greenish luminescent glow at night, a feature it uses to attract its prey. The rural Nepalese say that the dragons are rare, and give them a wide berth. No one in the group possessed a camera to record our encounter with the "dragon," so we had to make do with a detailed observation of its features before continuing on our way.[20]

Indian pythons are found in southern Nepal, though none officially attain the size of Poudal's creeping monster. The creature's lack of legs suggests a snake, and its 40-foot length is relatively modest when compared to other serpentine giants on file, but the huge mouth seems disproportionate to its reported size. Likewise, its glowing eyes (which none of Poudal's party noticed overnight) and its ability to suck in buffalos from yards away smack of mythology. In theory, its large diameter and sluggish movements might be explained by a recent large meal, but the question is moot. Thus far, Gibbons has not pursued his plan to find and film the Nepalese "dragon," nor have any further sightings surfaced.[21]

A disturbing series of pictures which appear to show a small boy having been eaten by a huge python

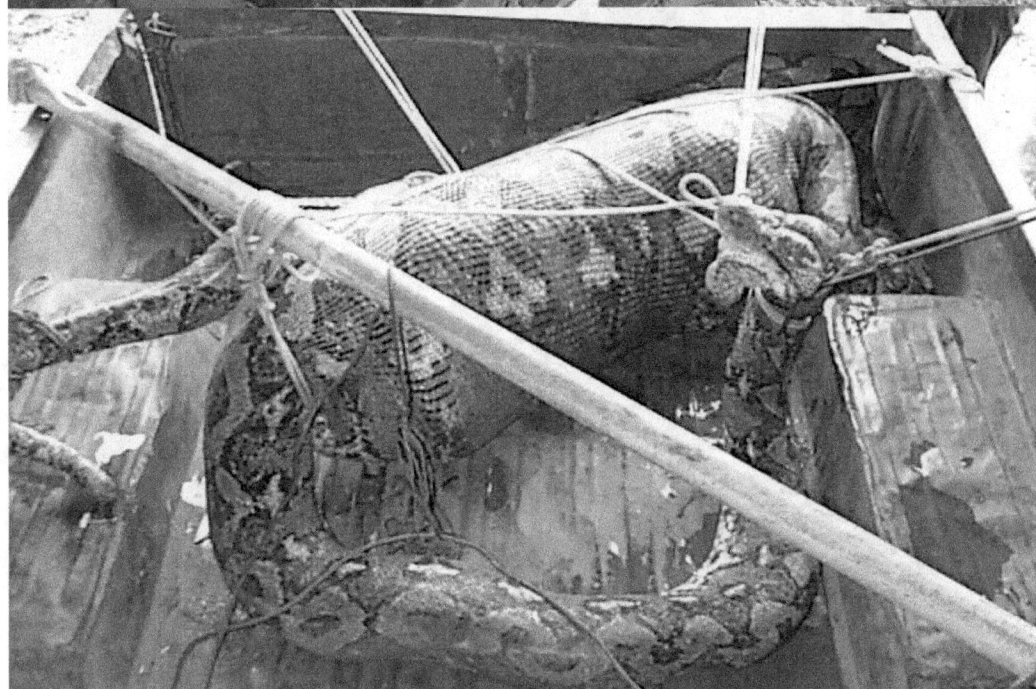

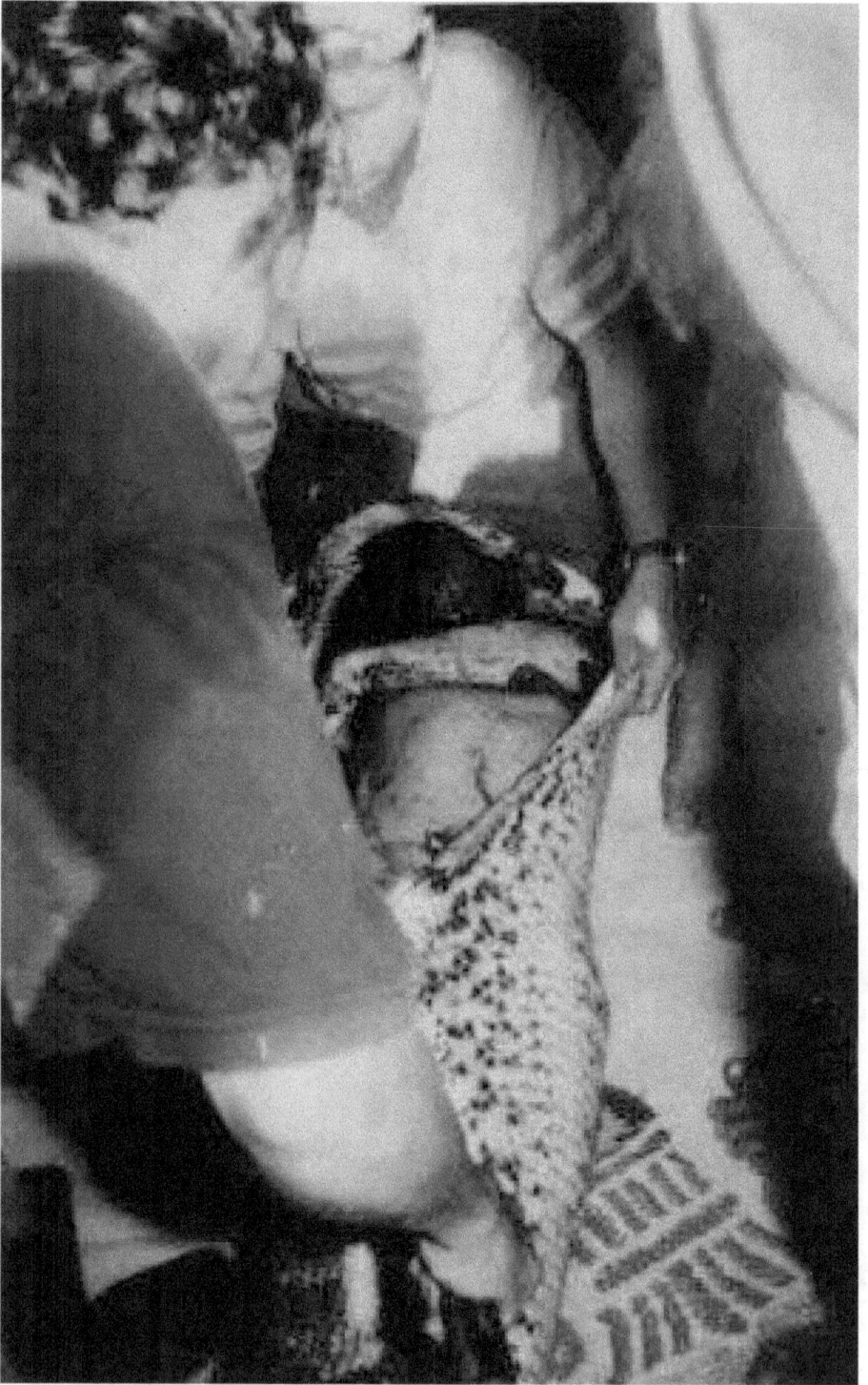

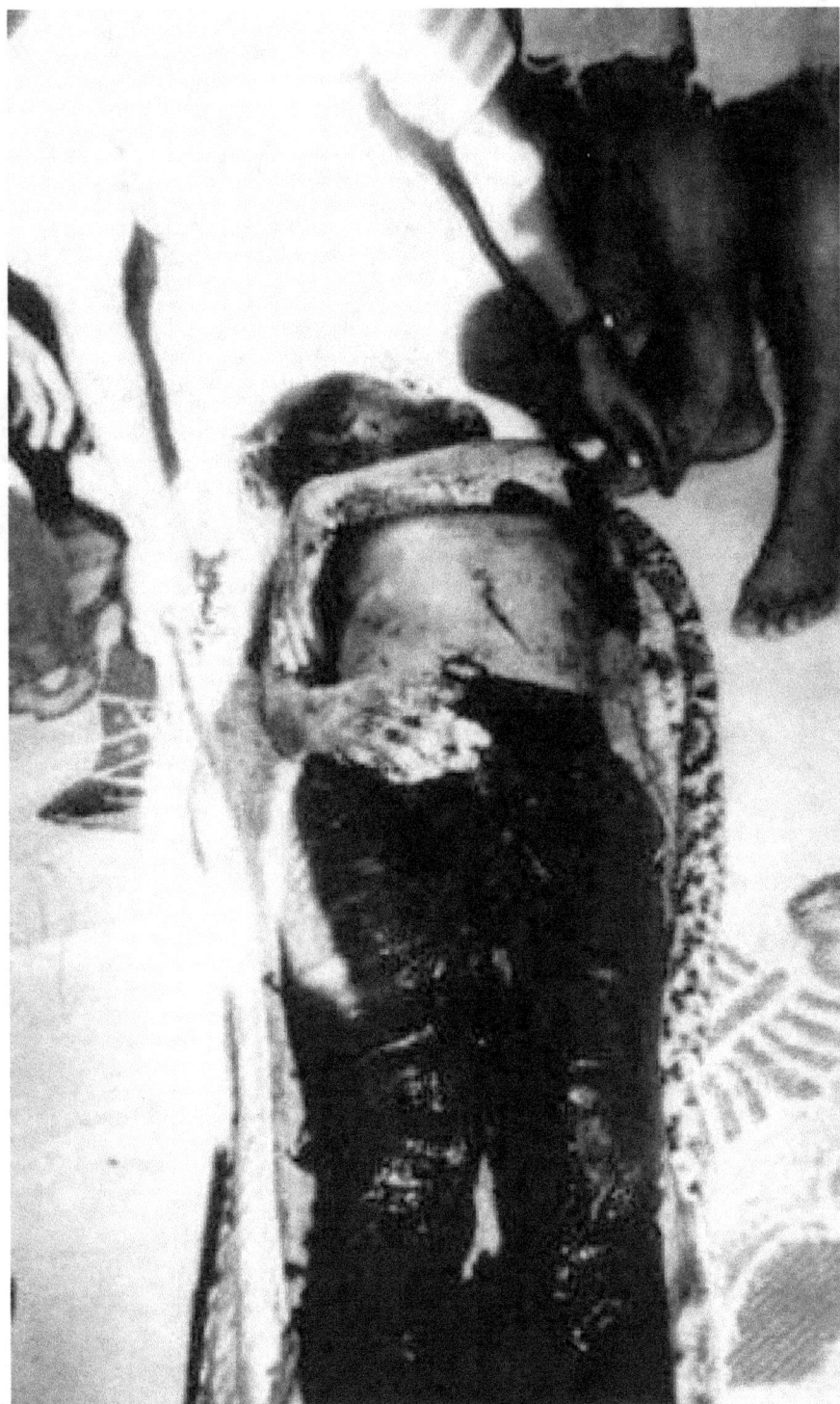

These pictures allegedly at least come from Borneo and show a
reticulated python.

However it has been alleged that the snake was in fact dead, and
had been cut open, so that a small boy could secrete himself inside for the purposes of providing an unpleasant series of photographs which could be posted on the internet horror-show websites such as rotten.com

We will probably never know the truth about them

India

Reports of giant snakes in India date back to Roman times. Pliny the Elder, in his *Natural History*, observed that :

> Africa produces elephants, but it is India that produces the largest, as well as the dragon, who is perpetually at war with the elephant, and is itself of so enormous a size, as easily to envelop the elephants with its folds, and encircle them in its coils. The contest is equally fatal to both; the elephant, vanquished, falls to the earth, and by its weight crushes the dragon which is entwined around it. The sagacity which every animal exhibits in its own behalf is wonderful, but in these it is remarkably so. The dragon has much difficulty in climbing up to so great a height, and therefore, watching the road, which bears marks of their footsteps, when going to feed, it darts down upon them from a lofty tree. The elephant knows that it is quite unable to struggle against the folds of the serpent, and so seeks for trees or rocks against which to rub itself.
>
> The dragon is on its guard against this, and tries to prevent it, by first of all confining the legs of the elephant with the folds of its tail; while the elephant, on the other hand, tries to disengage itself with its trunk. The dragon, however, thrusts its head into the nostrils, and thus, at the same moment, stops the breath, and wounds the most tender parts. When it is met unexpectedly, the dragon raises itself up, faces its opponent, and flies more especially at the eyes; this is the reason why elephants are so often found blind, and worn to a skeleton with hunger and misery.[22]

Aside from airborne attacks, Pliny reports that "dragons" also lie in rivers, waiting to snatch elephants as they approach the water. The reason, he claims, is that dragons crave the "remarkably cold" blood of elephants to quench their own thirst.[23] While Pliny's Indian "dragons" were large, exceeding 30 feet in length, he also cites a separate species of giant snake described by the Greek geographer Megasthenes (350-290 BCE), who served Syrian king Seleucus I as an ambassador to India during his latter years. Those serpents, Pliny claimed, "grow to such an immense size as to swallow stags and bulls."[24]

India's largest known serpent is *P. molurus,* with an official record length of 27 feet for a captive specimen in the United States, but larger individuals may yet exist. Conversely, if Pliny was correct in differentiating between Indian "dragons" and serpents, a still-unknown species may account for some of his reports. The tales of serpents blinding elephants suggests a spitting cobra, but no such species is recognized from the Indian subcontinent. Various species of spitting cobras *do* occur in China, Indochina, Indonesia, and Malaysia.[25]

Myanmar

This nation - formerly Burma - shares borders with Bangladesh, India, China, Laos and Cambodia. Its known reptilian fauna includes both reticulated and Burmese pythons (*Python molurus bivittatus*), as well as king cobras, but reports of an even larger unknown species persist.[26] Myanmar's mystery giant is the *bu-rin,* reported by villagers around Putao, in Kachin State (northern Myanmar). Putao, famous for its varied species of rare orchids, is inaccessible except by air during Myanmar's rainy season, and its surrounding jungles are poorly explored. Witnesses describe the *bu-rin* as an aquatic serpent 40-50 feet long, which attacks small boats and swimmers in all seasons.

Zoologist Alan Rabinowitz reports that "no one had firsthand knowledge of the creature, yet because of it, children were often discouraged from spending long periods of time in the water."[27]

Malaysia

Proceeding from Myanmar, southward to Peninsular Malaysia, we encounter more reports of giant snakes. In 1857 British naval officer Sherard Osborn (1822-75) published an account of Quedah - now Kedah, a state in northwestern Malaysia spanning 3,638 square miles - that included a report of huge serpents. He wrote: "The natives of Tamelan declared most of them to be of the boa-constrictor species, but spoke of monsters in the deep forests which might, if they came out, clear off the whole village. A pleasant feat, for which Jadie, with a wag of his sagacious head, assured me that an 'oular Bessar' or big snake was quite competent. It was strange but interesting to find amongst all Malays a strong belief in the extraordinary size to which the boa-constrictors or pythons would grow."[28]

Boa constrictors, as we know, are New World snakes, but Malaysia harbors reticulated and Burmese pythons, as well as the largest king cobras on record.[29] None match the dimensions of Kedah's *oular Bessar* - nor are reports of giant snakes restricted to the northern regions of Peninsular Malaysia.

Author Charles Gould cites an article from the *China Daily News* of 10 November 1880, reporting a desperate battle with a huge snake in the southern part of the peninsula. According to that article:

The *Straits Times* tells the following exciting python story:

"A sportsman, who a few days ago penetrated into the jungle lying between Buddoh and Sirangoon, came upon a lone hut in a district called Campong Batta, upon the roof of which the skin of an enormous boa or python (whichever may be the correct name) was spread out. The hut was occupied by a Malay and his wife, from whom our informant gathered the following extraordinary account. One night, about a week previously, the Malay was awakened by the cries of his wife for assistance. Being in perfect darkness, and supposing the alarm to be on account of thieves, he seized his sharp parang, and groped his way to her sleeping place, where his hand fell upon a slimy reptile. It was fully a minute before he could comprehend the entire situation, and when he did, he discovered that the whole of his wife's arm had been drawn down the monster's throat, whither the upper part of her body was slowly but surely following. Not daring to attack the monster at once for fear of causing his wife's death, the husband, with great presence of mind, seized two bags within reach, and commenced stuffing them into the corners of the snake's jaws, by means of which he succeeded in forcing them wider and releasing his wife's arm. No sooner had the boa lost his prey than he attacked the husband, whom he began encircling with his fatal coils; but holding out both arms, and watching his opportunity, he attacked the monster so vigorously with his parang that it suddenly unwound itself and vanished through the opening beneath the attap sides of the hut. His clothes were covered with blood, as was also the floor of the hut, and his wife's arm was blue with the squeezing it received between the boa's jaws. At daylight the husband discovered his patch of plantain trees nearly ruined, where the boa, writhing in agony, had broke off the trees at the roots, and in the midst of the debris lay the monster itself, dead. The Malay assured our informant that he had received no less than sixty dollars from Chinese, who came from long distances to purchase pieces of the flesh on account of its supposed medical properties, and that he had refused six dollars for the skin, which he preferred to retain as a trophy. It was greatly decomposed, having been some days exposed in the open air, and useless for curing. There is no telling what may have been the measurement of this large reptile, but the skin, probably greatly stretched by unskilful [sic] removal, measured between seven and eight fathoms [42-48 feet]." [30]

More recent accounts of reptilian giants come from the Tasik Bera, a system of freshwater lakes and swamps spanning some 207 square miles in southwestern Pahang, the third-largest state of Peninsular Malaysia. The Tasik Bera's known wildlife includes 200 bird species, 50 mammals, and more than 95 species of fish. Native creatures include elephants, tigers, clouded leopards, tapirs, and other endangered species - plus, perhaps, something else.[31]

British journalist Stewart Wavell toured Malaysia in 1951 and pursued the legend of a huge snake dwelling in the Tasik Bera region, known to native Semelai tribesmen as *ular tedong* or *nagaq*. They considered the *nagaq* to be a giant cobra, and distinguished it specifically by name from the *nāga*, a legendary aquatic serpent or dragon of Buddhist and Hindu mythology. Wavell never spotted the monster himself, but he interviewed an eyewitness - the police chief of Negri Sembilan - who had seen a silver-gray beast raise its neck 15 feet above the Tasik Bera's surface. The beast's body, he said, included "two upper contour curves" visible above water. Wavell concluded that "[t]he officer is utterly reliable and has a fine record. He doesn't exaggerate normally and I should accept his word every time. But as for the Monster, I must say it is a bit of a queer story."[32]

Thus encouraged, Wavell returned to the Tasik Bera and toured the Pahang River with native guides, reaching another swampy lake called Sungei Bera. There, he collected more Semelai tales of the *nagaq*, including descriptions of its booming cry which seemed unlike any sound made by snakes known to science. Armed with both a camera and tape recorder, Wavell stalked the creatures again, but failed to set eyes upon one. His sole reward was "[a] single staccato cry from the middle of the lake [that] chilled my blood with fear. It was a snort: more like a bellow - shrill and strident like a ship's horn, an elephant's trumpet, and a sea-lion's bark all rolled into one. I was momentarily petrified then frantically switched on the recorder and waited for the next cry - but it never came."[33]

In 1979 a huge snake appeared at Semenyih, a small town in the western Malaysian state of Selangor. It dwelt in an abandoned mining pool, appearing first to fisherman Lebai Ramli, who summoned other witnesses. According to the villagers' reports, the snake's head was the size of a scooter wheel or oil drum, while its body resembled a floating log. Locals blamed the serpent for the disappearances of two buffaloes and other livestock, while resident Enick Jaafar suddenly recalled sightings of a similar reptile soon after World War II.[34]

Singapore

This island nation at the southern tip of the Malay Peninsula is one of Earth's few remaining city-states, and at 272 square miles it ranks as the smallest country in Southeast Asia.[35] It is not so small, however, that it has no tales of giant snakes to offer.

British civil servant John McNair (1829-1910) joined the British East India Company at age 17 and served initially in India, then transferred to the Straits Settlements (Penang, Malacca and Singapore) in 1857. Twenty years later, after serving in various government posts, he was appointed Resident Councillor to the Governor of Penang, maintaining that position until poor health forced his resignation in 1884.[36] Meanwhile, in 1878, he published a book titled *Perak and the Malays*, which includes the following passage:

One of the keenest sportsmen in Singapore gives an account of a monster that he encountered. He had wounded a boar in the jungle, and was following its tracks with his dogs, when, in penetrating further into the forest, he found the dogs at bay, and, advancing cautiously, prepared for another shot at the boar; to his surprise, however, he found that the dogs were baying a huge python, which had seized the boar, thrown its coils around the unfortunate beast, and was crushing it to death. A well-directed shot laid the reptile writhing upon the ground, and it proved to be about thirty feet long. But such instances of extreme length are rare.[37]

Rare enough, in fact, that modern science has yet to confirm a single snake of any species reaching 30 feet, despite the numerous reports of larger specimens.

The *Nāga*

Huge serpents or dragons imbued with supernatural powers are said to inhabit Thailand and the nations of former French Indochina, including Cambodia, Laos and Vietnam. Collectively known as *nāga*, they reportedly inhabit caves and rivers - including the mighty Mekong, which flows from northwestern China's Qinghai Province through Tibet, dividing Laos from Myanmar, Thailand and Cambodia, spilling at last into Vietnam's Mekong Delta.[38]

Standard descriptions of the *nāga* depict a black serpent ranging up to 60 feet long, often with a crest atop its head, and sometimes with a jewel embedded in its skull. When confronting enemies or prey, the *nāga* raises its head like a cobra. Conflicting stories endow it with venomous fangs or an ability to spit venom as some cobras do, while others describe *nāgas* crushing prey in their coils. Reports of *nāga* killing from long-distance, with a glance, clearly owe more to legend (like the basilisk of Medieval Europe) than to any flesh-and-blood reptile.[39]

Another strange trait of the *nāga* is its supposed ability to launch crimson fireballs. Some legends claim these are the *nāga*'s floating eggs, while others treat them as a supernatural display akin to fire-breathing by certain European dragons. The displays appear most often in October, coinciding with a full moon at the close of Buddhist Lent, and Vietnamese believers gather on the Mekong's banks in tens of thousands, to view the spectacle each 13 October. CFZ investigator Richard Freeman observed the ritual in October 2000 and concluded that the fireballs were "nothing more than fireworks, the relatively noiseless kind that fade away rather than exploding, much like maritime distress flares." Apparent validation of that judgment came three years later, from Laos, when officials of the Lao National Authority and Vientiane Municipality's Pak Ngum district scheduled a *nāga* fireball display for 200,000 paying tourists. Curiously, one week later, Thailand's Science Ministry announced that the *nāga* lights were a natural phenomenon, caused by flammable phosphine gas rising from the Mekong.[40] With or without fireballs, stories persist of giant-snake sightings in Southeast Asia. Richard Freeman collected several eyewitness reports during his October 2000 expedition, including the following:

- In May 1980 fisherman Prancha Pongpaew found 17 large eggs in northern Thailand's River Ping. Each was the size of a watermelon, and they "seemed to be linked." Prancha brought five back to his home in Songtham village, where they were opened, emitting foul odors. Later that night, wailing sounds woke Songtham's residents and they beheld two black serpents as large as palm trees, raising their crested heads from the nearby river. Religious ceremonies banished the monsters, which never returned. Freeman accurately notes that known reptiles do not lay linked eggs in water, yet his conclusion that "the nagas and the eggs were two unconnected events" strains the bounds of coincidence.[41]
- In 1990 a certain Mr. Pimpa met a *nāga* on dry land, while exploring a jungle cave. He did not

see the whole reptile, but estimated that its visible body measured 65 feet in length and was 30-35 inches thick. Mr. Pimpa led Freeman to the *nāga*'s lair, but it was not at home. Freeman found the tunnel half-filled with water, obstructed by sharp stalactites, the walls coated with "primal slime" - but no proof of a giant snake's presence.[42]

- In 1992 workmen gathered to demolish an old temple at Phon Pisai and erect a new one, but they were prevented from proceeding by a huge black snake that occupied the crumbling structure. Every time the crew approached, the reptile showed its head and several yards of body, while most of its bulk remained hidden from view. Witnesses including laborers, monks, and the temple's abbot described the snake as "very thick," but were unable to judge its length. The monster vanished one night, placated by ritual offerings.[43]

- In 1995, while approaching the town of Nongkhai on a bus, witness Malinee Phisaphan and a companion saw a 17-foot snake swimming beneath a highway bridge. Its size was well within the range of either *P. molurus* or *P. reticulatus,* but neither species matches the mystery snake's jet-black colour. It may have been a melanistic specimen, as Freeman suggests, or it may represent a new species.[44]

- In 1997 the *nāga* returned to Phon Pisai, revealing itself to 30 witnesses, including Chief of Police Suphat. The observers were hiking on cliffs above the Mekong when they saw the monstrous black serpent swimming below. They estimated its length at a whopping 230 feet and described its horizontal undulations, characteristic of a reptile or fish. Freeman concluded that Chief Suphat and his friends had actually seen "several nagas swimming in line, perhaps males in pursuit of a female."[45]

While in Thailand, Freeman also examined supposed *nāga* bones (which proved to be common elephant's teeth) and viewed a film of a *nāga* swimming in the Mekong (dismissed in turn as a "wobbly badly filmed log being bobbed up and down by the current"). Despite those disappointments, and a preconception that *nāgas* might be large estuarine crocodiles (*Crocodylus porosus*), Freeman ultimately "came away convinced in the existence of some kind of huge reptile unknown to science lurking in the primal morasses of Indochina." Specifically, he suggests a relict form of *Madtsoiidae*, primitive snakes whose fossil record extends from the Upper Cretaceous Period to the Pleistocene. Matsoiid species included *Gigantophis garstini,* the largest snake acknowledged by science at 35 feet, presumed extinct for 40 million years.[46]

A giant-snake report from Vietnam, dating from 1964, sadly lacks anything resembling documentation. In fact, it sounds like a garbled and exaggerated version of Remy Van Lierde's Congo adventure from 1959. Nonetheless, I present it here for whatever it may be worth. Anonymous Internet poster "Chunga" writes [grammar and spelling uncorrected]:

> Once I saw a documentary of Vietnam photographer's. One of the army photog's spoke of going out over a particular area of the Vietnamese jungle to photograph roads and pathways for map making. He said that while flying over a certain little winding dirt road way out in nowhere while hanging out the open sidedoor while strapped on he noticed a large grey/green rope like image crossing over the road from one side into the jungle growth on the other side. He instructed the pilot to get lower over that area and when they dropped down they could both see that the thing across the road was moving. They stayed in the area and he photographed the large tubelike thing for somethime, then suddenly out of the jungle tree tops rose a large snake head and quite a bit of it's upper body. They were freaked out, it rose up at least 200 feet from the tree tops and had its head curled so it was even with the ground. Just swaying and looking at them. The helicopter was about 100 feet from the head. Then it moved in a way that made the photographer feel that it was

going to try to strike at the helicopter. He could not believe this, it felt strong enough to attempt this with a noisy moving thing like a freakin helicopter. He said the head was the size of a very large horse and its eyes had to be 3 feet in diameter. They flew away for a few minutes trying to digest the whole thing and talked about what to do. The photographer took several pictures of the road image and of the head and neck out of the treetops. They showed these and I remember seeing blurry, hazey black and white images of what looked very much like a gigantic snake. They flew back to the area a few minutes later and it was no where. He said when he got back to base he told about it and the pilot did too. They talked to native area people and they were told that these snakes are rare, but have been encountered by tribal family members for 100s of years and they believe these snakes to be very old. They are known by the locals as The Bull Eaters. I dont know the name of the photographer or the area of Vietnam they were in, I cant remember the documentary name or the channel it was on. Does anyone else know of this incident or the documentary name? I have looked for it in books, online and by word of mouth with Vietnamese merchants and families in my area. No Luck. Did you see these photo's?[47]

I had no better luck than "Chunga" in pursuing this report, which may be wholly fabricated.

Two years after the supposed Vietnam incident, another clutch of *nāga* eggs was found, this time in Laos. While excavating the Mekong's muddy bank in 1966, near the Lan Xang Hotel in the Laotian capital of Vientiane, an unnamed peasant allegedly found three huge white eggs. A *nāga* soon appeared to him in dreams, demanding return of the eggs, threatening a flood if they were not delivered, but the peasant gave them instead to Kong Le, chief of staff for the Laotian army from 12 February 1963 to 7 October 1966 (when he fled the country). Kong Le reportedly showed the eggs to Prince Souvanna Phouma, the Laotian Prime Minister, who rejected Kong Le's bid for a religious ceremony of atonement. Vientiane suffered catastrophic flooding during August and September 1966, whereupon the eggs were lost. While Kong Le presently resides as an exile in the United States, I was unable to locate him for an interview.[48]

Indonesia

Published reports of giant snakes in Indonesia date from 1658, when Gulielmus Piso posthumously published *Historiae Naturalis et Medicae Indiae Orientalis,* by Dr. Jacob Bontius (1592-1631). Writing of pythons in the Asiatic Isles (later the Dutch East Indies), Bontius said, "[T]he great ones sometimes exceed thirty-six feet; and have such capacity of throat and stomach that they swallow whole boars."[49]

Nearly 200 years later, in 1852, British naturalist and lawyer William Broderip published his *Leaves from the Note-book of a Naturalist,* including the description of a 30-foot snake that attacked the crew of a Malay proa anchored off the island of Celebes (now Sulawesi in the Sunda Islands, between Borneo and the Malukus). In 1857, Sherard Osborn wrote: "[The Malays] all maintained that in the secluded forests of Sumatra and Borneo, as well as on some of the smaller islands which were not inhabited, these snakes were occasionally found of forty or fifty feet in length."[50]

Cuthbert Collingwood filed a similar hearsay report in 1868, writing that "Mr. Low assured me that he had seen one [python] killed measuring twenty-six feet, and I heard on good authority of one of twenty-nine feet being killed [on Celebes]. In Borneo they were said to attain forty feet, but for this I cannot vouch."[51]

On 21 May, 1877, crewmen aboard the barque *Georgina* reported a sea monster sighting from the Indian

Ocean, west of Sumatra. They described the beast as "a large serpent about forty or fifty feet long, grey and yellow in colour, and ten or eleven inches thick." Thirty years later, in summer 1907, witness S. Clayton, third officer of the China Navigations Company's ship *Taiyuan,* saw an even larger reptile in the Celebes Sea.[52] After observing the beast through binoculars, he wrote:

> Almost abreast now and no longer foreshortened, was stretched an enormous writhing serpent of fabulous size. Yet monstrous as it was, its proportions were as fine as our English Grass Snake, though the head may have been more angular and more boldly outlined. The creature so far as I could see, appeared to be a perfect replica of the land snake. It was at least seventy feet long, having a girth corresponding in size to its length for a snake. It was a rather dark cane colour (of course, I saw none of the underside) having uneven dark brown patches or figuring such as one might expect to find joined by concatention on closer inspection. Its convolutions were not vertical as many illustrations used to depict them, but horizontal in the plane of the surface of the water: the serpent being just submerged only. Apart from its writhing motion I could gather nothing about its propulsion....It did not seem to panic at the very close proximity of the ship; but continued steadily along our side, its course being exactly opposite to ours. I have wondered if it could have been a python of staggering dimensions, something on the lines of the S. American Anaconda, but far larger, crossing between the islands. Indeed my private opinion of the serpent's length has always been eighty feet, but I state it here as seventy purposely to err on the side opposed to exaggeration.[53]

Bernard Heuvelmans accepted that judgment from Clayton, reporting "no doubt that it was a snake, perhaps some huge python," but rejected a similar verdict for the *Georgina* monster, ranking it as a form of unknown aquatic cryptid which Heuvelmans designated the "yellow-belly" or the "super-eel."[54]

The Indonesian archipelago includes 18,000 known islands. The official count was 17,000 until February 2003, when satellite photographs revealed an additional 1,000 previously uncharted islands. Humans inhabit only one-third of Indonesia's islands, with more than half of the country's population residing on Java. Clearly, there is much ground still to be explored, with a potential for revealing new species.[55]

Even on Borneo - third-largest island in the world at 287,000 square miles, with a population of 16 million in 2000 - zoological secrets await. Between 1996 and 2007, more than 400 new animal species were identified on Borneo, capped by revelation of a new clouded leopard species in March 2007.[56] We should, perhaps, remain a bit more open-minded with regard to giant-snake reports from this vicinity.

The CFZ's Richard Freeman made two snake-hunting trips to nearby Sumatra - 181,467 square miles, 45 million inhabitants - during 2003-04. On the latter expedition, Freeman questioned members of the jungle-dwelling Kubu tribe concerning their encounters with large pythons. No specimens were seen, but the tribe's chief described recent captures of pythons 23, 26 and 33 feet long. Three Kubu tribesmen also related close-range encounters with a different species of serpent, averaging 33 feet, which sported cow-like horns atop the head. At maturity, the Kubu claimed, these snakes mimicked the Chinese dragons by sprouting four legs and assuming the guise of a huge crocodile. Freeman found that tale of transformation unique to the Kubu, while noting that their name for the horned crocodiles was *naga,* bearing no physical resemblance to the giant snakes of Indochina.[57]

One final mystery involves the fabled island of Zanedj or Zanig, described by Arab historian and geographer Abu al-Hasan Ali ibn al-Husayn íbn Ali al-Mas'udi (896-956). Vaguely described as "a large island on the confines of China toward India," Zanedj boasted a mountain called "Nacan" whose serpents were large enough to swallow oxen, buffaloes and elephants.

French author Pierre Amédée Jaubert, writing in 1836, speculated that Mount Nacan might be identical with Kini-Balu, a 12,000-foot peak on Borneo's northern coast, while other sources suggest that the neighboring island of Taprobana - also described by al-Mas'udi - may be Sri Lanka. Meanwhile, the map of Al Idrisi, drawn in 1150, locates multiple islands of "Zaledj" or "Zanedj" near Zanzibar, off the eastern coast of Africa. That riddle, like the identity of the island's titanic serpents, likely will never be resolved.[58]

The Philippines

Our final stop in Asia takes us to the Philippines, known from their "discovery" in 1565 until their "liberation" by America in 1898 as the Spanish East Indies. The Philippines comprise an archipelago of 7,107 islands, bordering the nations of Indonesia, Malaysia, Palau and China in the Western Pacific, but sharing no land borders with any neighbours.[59]

Alfred Russell Wallace (1823-1913), a British anthropologist, biologist, explorer and geographer, referred briefly to giant Filipino pythons in his book *Australasia,* published in 1883. Without providing details, Wallace wrote that certain snakes exceeded 40 feet in length and dined upon "young cattle." Charles Gould, writing in 1886, chastised Wallace for merely reporting hearsay, yet the fact remains that portions of the Philippines are still unsettled and effectively unexplored. Captain Fumio Nakahira, a Japanese "holdout" veteran of World War II, successfully concealed himself on Mindoro - the nation's seventh-largest island, with a population exceeding 1 million - from April 1945 until April 1980.[60]

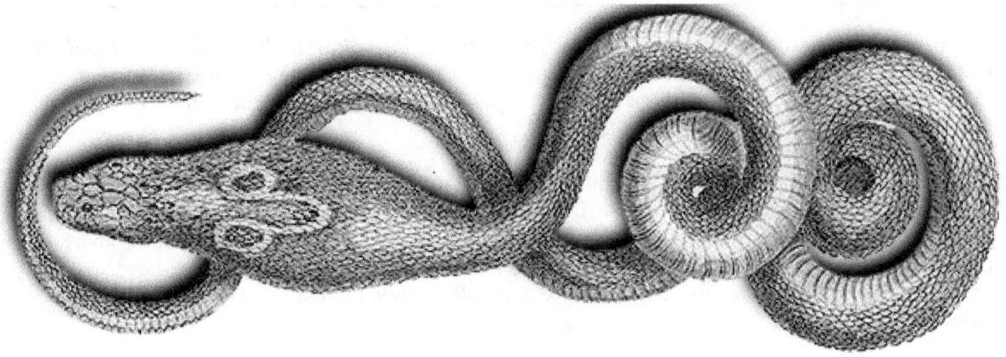

Chapter 5: Oceania

O ceania, broadly defined, includes Australia and the myriad island nations of Melanesia, Micronesia, and Polynesia - in short, the bulk of the western and southern Pacific Ocean. Melanesia extends from the western side of the West Pacific to the Arafura Sea, north and northeast of Australia. It includes New Guinea and the Bismarck Archipelago - Bougainville, Buka, New Britain, New Ireland, and 12 other island groups - plus New Caledonia, the Solomons, Fiji, Norfolk Island, Vanuatu, the Malukus, and Torres Strait Islands. Micronesia includes the several hundred islands in the Federated States of Micronesia, plus Wake Island and other famous battlegrounds of World War II in the Mariana, Marshal and Palau island groups. Polynesia includes New Zealand and more than 1,000 other islands scattered over the central and southern Pacific.[1]

That said, most of Oceania's vast realm does not concern us in our quest for giant snakes. New Zealand claims no indigenous snakes, and known python species inhabit only a small percentage of the Pacific islands. Neither reticulated pythons nor *Python molurus* are acknowledged anywhere in Oceania.[2]

Australia

Continents are arbitrarily defined by size, with Australia ranked as the smallest, at 2.9 million square miles. It is thus three and a half times larger than Earth's largest island - Greenland - but the line blurs for some, who persist in calling Australia "the Island Continent." Science recognizes more than 170 species of serpents Down Under, including 100 that are venomous.[3]

Among those snakes are 10 species and subspecies of python, including one of the Big Six. The amethystine python (*Morelia amethistina*) is Australia's largest known species, with various sources disputing its maximum size. Clifford Pope claims "ample evidence" of amethystine pythons reaching 20 feet, and cites a dead specimen measured by one S. Dean in 1954 at 23 feet 8 inches. Louis Robichaux described a larger specimen, ranging in various still-unconfirmed reports from 25 to 28 feet long. Internet sources claim that amethystine pythons have been measured at 27 feet 7 inches, but call that size "quite exceptional" and peg "large" specimens around 18 feet.[4]

Three other species of *Morelia* are also found in Australia, including Bredl's python (*M. bredli*) and the Oenpelli python (*M. oenpelliensis*) in the Northern Territory, and the rough-scaled python (*M. carinata*) in Western Australia. The largest species, *M. oenpelliensis,* holds a record length of 13 feet, while the others generally range from 6 to 8 feet long.[5]

Two other python genera inhabit Australia, including six recognized species. The genus *Antaresia* includes the children's python (*A. childreni*), with a record length of 3 feet 7 inches; the pygmy python (*A. perthensis*), at 28 inches maximum; the spotted python (*A. maculosa*), 4 feet 5 inches; and Stimson's python (*A. stimsoni*), 3 feet 3 inches. In the genus *Aspidites,* we find the black-headed python (*A. melanocephalus*), with a record length of 9 feet, and Ramsay's python (*A. ramsayi*), at 7 feet 6 inches. None qualify as giants.[6]

As on other continents, of course, native mythology and latter-day reports allude to serpents far surpassing any known species in size. Australia's archetypal monster snake is the rainbow serpent, depicted in aboriginal rock art dating from 4000 BCE. Various tribes knew the titanic reptile by different names, including *Almudj, Bobi-Bobi, Dhakkan, Galeru, Ganba, Julunggul, Kalseru, Karia, Kinmanggur, Langal, Mindi, Muit, Ngalbjod, Taipan, Takkan, Ungur, Wagyl, Woinunggur, Wollunqua, Wollunquain, Wonambi, Worombi, Yero, Yingarna,* and *Yurlunggur.*[7]

Today, the taipan name applies to any of six serpents in the genus *Oxyuranus,* which - though hardly giants, with a record length of 12 feet - rank among the world's most venomous. One bite from *O. microlepidotus* delivers sufficient venom to kill 100 adult humans or 250,000 mice. Rather curiously, authors Christopher Chippendale, Paul Tacon, and Meredith Wilson concluded a 1996 statistical analysis of 107 rock-art samples from Arnhem Land (Northern Territory) by suggesting that the rainbow serpent was modeled on a fish, the ribboned pipefish (*Haliichythys taeniophora*), which does not exceed a foot in length.[8]

Meanwhile, some evidence suggests that specimens of *Mindi* may still survive along South Australia's Lower Murray River - or did, at least, through the mid-19th century. Charles Gould, writing in 1896, related conversations with one G.R. Moffat, who lived and worked along the Lower Murray, between Swan Hill and Darling Junction, during 1857-67. In that era, "black fellows" of the Wemba tribe described encounters with the *Mindi,* also called the Mallee snake, after a species of eucalyptus tree found in the region. Witnesses described the snakes as 30-40 feet long, known for their "great girth, swiftness, and intensely disgusting odor." The reptiles lurked at waterholes or hung from tree branches to ambush emus.[9]

Moffat referred Gould to a white eyewitness, Peter Beveridge of Swan Hill Station, who confirmed the *Mindi's* general description and supplied a clipping from the *Proceedings of the Royal Society of Tasmania,* dated 13 September 1880.[10] The brief article read:

Mr. C.M. Officer states -

> "With reference to the Mindi or Mallee snake, it has often been described to me as a formidable creature of at least 30 feet in length, which confined itself to the Mallee scrub. No one, however, has ever seen one, for the simple reason that to see it is to die, so fierce it is, and so great its power of destruction. Like the Bunyip, I believe the Mindi to be a myth, a mere tradition."[11]

Perhaps, yet Darling River resident Henry Lidell furnished Gould with reports from local ranchers, logged in 1871-72, concerning "the existence of large serpents of the boa species in an adjacent locality," east of the Darling and Murray Rivers, around Pooncaira, New South Wales. Those witnesses described the snake "as being numerous, in barren and rocky places, among big boulders; fully 40 feet long; as thick as a man's thigh; and having the same remarkable odour described by Mr. Moffat. They spoke of them as quite common, and not at all phenomenal, between Wentworth and Pooncaira."[12]

New Guinea

New Guinea is Earth's second-largest island (after Greenland), at 303,381 square miles. Thanks to its tallest mountain - Puncak Jaya, also called Mount Carstensz - it also ranks as the worlds fourth-highest landmass. Politics divides the island, with its western half comprising the Indonesian provinces of Papua and West Papua, while the eastern half forms the independent nation of Papua New Guinea. An estimated 7.1 million inhabitants speak 1,073 known languages, but others may yet remain undiscovered. The Dani tribe was "lost" until 1938, while the Yali met their first white explorer in 1976. Reports of large prehistoric reptiles at large on New Guinea continue in the 21st century.[13]

New Guinea has its share of pythons, including the amethystine and three other species from the same genus. They include Boelen's python (*M. boeleni*), with a record length of 8 feet; the carpet python (*M. spilota*), 12 feet maximum; and the green tree python (*M. viridis*), topped at 7 feet. Two other monotypic genera (having a single species only) also inhabit New Guinea. The Papuan python (*Apodora papuana*) may reach an impressive 17 feet, while the much smaller D'Albert's water python (*Leiopython albertisii*) does not officially exceed 7 feet.[14]

My father served as a member of the U.S. Army in New Guinea, during World War II, and later shared with me the story of his own encounter with a large python. His anti- aircraft unit was in transit, riding open trucks over a jungle road at night, when the lead driver stopped abruptly. He had seen a "log" lying across the road, which proved to be a python crossing at its leisure. I cannot surmise the reptile's length from what my father said, long afterward - except to note that the driver feared to run over the snake, believing it might lash out and encircle his truck, striking at the soldiers seated in its open bed. The small convoy was stalled for several minutes while the snake proceeded on its way.

The standard U.S. Army truck from 1941 onward was the Dodge WC62, which measured 18 feet 9 inches long, 6 feet 3 inches wide, and 7 feet 3 inches high. A larger model, the Diamond T968, measured 22 feet 4.5 inches long, 8 feet wide, and 9 feet 10.5 inches tall.[15] In either case, a snake long enough to encircle the truck with a single coil would range from 27-35 feet long - but we must grant the possiblity of fear-inspired exaggeration on the driver's part.

Island Giants

Moving beyond New Guinea, through the scattered island realms of Oceania, we find pythons in some locales, while others harbour none. The largest of the lot, *Morelia amethistina*, ranges through the Bismarck Archipelago to Umboi Island (between Papua New Guinea and New Britain), Misool and Salawati (in the Raja Ampat Islands), the Kiriwina Islands, D'Entrecasteaux Islands, and the Louisiade Archipelago (all surrounding New Guinea in the Solomon Sea). *M. boelieni* also inhabits Goodenough Island (east of New Guinea), while *M. viridis* inhabits the D'Entrecasteaux Islands and Normanby Island, farther to the south.[16]

Apadora papuana, with a record length of 17 feet, may be found on various islands claimed by New Guinea, from Misool in the Raja Ampat chain to Fergusson Island, largest of the D'Entrecasteaux chain. Another monotypic python genus, *Bothrochilus*, fields its only species (*B. boa*) in the Bismarck Archipelago, with specimens reported from Umboi, New Britain, Gasmata, Duke of York, Mioko, New Ireland, Tatau, the New Hanover Islands, and Nissan Island. The largest representative of *Bothrochilus* on record measured six feet in length.[17]

Rather surprisingly, the vast Pacific's only giant snake of record seems to be the *amam,* described by Irish author Ronan Coghlan as a huge sea snake from Micronesian mythology, so large that "you can sail into it, only realising what you have done when you can no longer see the stars." Curiously, Coghlan's source - the Internet website *Encyclopedia Mythica* - says no such thing. Its brief entry on the *amam* describes a black snake "as big as an island house," inhabiting the waters between Mili Atoll in the Marshall Islands and Kiribati (formerly the Gilbert Islands, plus the Line and Phoenix Islands).[18]

The Boas

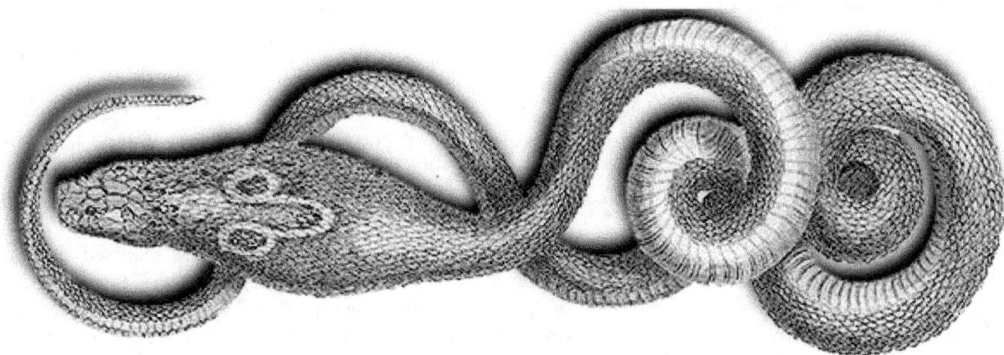

Chapter 6: South America

South America is Earth's fourth-largest continent, spanning 6,890,000 square miles. It presently includes 12 independent nations and three dependencies, ruled from afar by Great Britain and France. An estimated 382 million persons inhabit South America, for a supposed population density of 55 persons per square mile, but the reality is very different.[1] In fact, the continent's human population is distributed as follows:

- *Argentina* has 40.3 million recognized citizens. Of those, 15.5 million live in greater Buenos Aires, while 10.4 million occupy 16 other large cities. Nationwide, 123 more cities claim populations of 45,000 to 150,000 each.[2]
- *Bolivia* has 9.1 million inhabitants, 1.2 million of whom live in Santa Cruz de la Sierra. Nine other cities house a total of 3.3 million.[3]
- *Brazil* claimed nearly 184 million residents in 2007. Of those, 14.1 million lived in Porto Alegre, 11.4 million lived in Rio de Janeiro, and 10 million lived in São Paulo. Twenty-six other cities claimed 40.8 million.[4]
- *Chile,* with an estimated 16.6 million inhabitants, finds nearly one-third of that total (5.4 million) confined to Santiago. Twenty-nine other metropolitan centers claim 7.9 million residents.[5]
- *Colombia* claimed a population of 44,087,000 in April 2008. Of those, 11,903,661 lived in the three principal cities of Bogotá, Cali, and Medellin. Thirty-seven smaller cities harbored anoter 12.4 million.[6]
- *Ecuador* estimates its population at 13.8 million souls. Guayquill claims 2.2 million of those, with another 1.5 million in Quito, and eight other cities house 1.7 million.[7]
- *Guyana* reports 751,000 inhabitants, of whom 28 percent - some 215,000 - reside in the capital at Georgetown.[8]
- *Paraguay,* with 6.2 million residents, finds one in every five (1.3 million) living in Asunción.[9]
- *Peru* claims 28.7 million citizens. Nearly one-third of those (7.9 million) live in Lima, while 19 other sizeable cities claim a total of 6.5 million.[10]
- *Suriname* had an estimated population of 470,784 in 2005. More than half of those, some

250,000, reside in the capital at Paramaribo.[11]

- *Uruguay* boasts a population of 3.5 million, with 1.3 million living in Montevideo. Thirty-eight other cities and towns house another 1.2 million.[12]
- *Venezuela* is larger, with an estimated population of 28.2 million in 2008. Of those, 3.3 million inhabit Caracas, 2.06 million live in Maracaibo, and 1.4 million reside in Valencia. Thirty-seven other cities with six-figure head counts contain 9.2 million inhabitants.[13]

South America's dominant geographical features are the Andes (Earth's longest exposed mountain range) and the Amazon Basin - the planet's largest rain forest, sprawling over 3,178,876 square miles. The snow-capped Andes do not feature in our search for giant snakes, but the Amazon Basin has produced some of the most dramatic sightings on record. Brazil claims 54 percent of the Amazon, with the remainder shared by Guyana, Venezuela, Colombia, Peru, and Bolivia. An estimated 26 million persons inhabit the region, of which 11 million are Brazilian. The basin's largest cities are also Brazilian, with Manaus claiming 2.8 million against Belém's 2.1 million. The next-largest Amazonian city - Iquitos, in Peru - claims only 400,000 inhabitants.[14]

And in the jungle's depths dwell giants.

Boa Constrictors

All boas kill by constricting their prey, but only one species is properly known as the "boa constrictor." Ten subspecies confuse the issue, but the classic boa constrictor - also known as the red-tailed boa, or *macajuel* (in Trinidad) - is *Boa constrictor constrictor.*[15]

By all accounts, boa constrictors rank as the smallest of the "Big Six" serpents, but their maximum size is disputed. Clifford Pope acknowledges a record length of "18 or 19 feet," specifically referring to a specimen shot by Colin Pittendrigh on Trinidad, which measured 18.5 feet. Pope also notes that German author Robert Mertens "heard of" a 19.5-foot boa in El Salvador, but questioned the validity of the report. Apparently, the largest confirmed living specimen is a 15-footer from Suriname, housed at the San Diego Zoo.[16]

One report of a substantially larger boa constrictor appears in John Browne's *An Affecting Narrative of the Extraordinary Adventures and Sufferings of Six Deserters from the Artillery of the Garrison of St. Helena in the Year 1799,* published in London during 1802. The fugitives in question make their way to South America and plunge into its jungle, where, among other perils they face a presumed boa constrictor (called *ibibaboka* in the narrative) that measures 29 feet 3 inches long.[17] Scottish explorer George Gardner described even larger boas in his 1846 tome, *Travels in the Interior of Brazil.* The largest, he said, inhabited the Province of Goyaz (now the state of Goiás, in central Brazil). Gardner wrote:

> The largest I ever saw was at this place, but it was not alive. Some weeks before our arrival at Sapé, the favourite riding horse of Sanhor Lagoeira, which had been put out to pasture not far from the house, could not be found, although strict search was made for it all over the Fazenda. Shortly after this, one of his vaqueros, in going through a wood by the side of a small river, saw an enormous Boa, suspended in the fork of a tree which hung over the water; it was dead, but had evidently been floated down alive by a recent flood, and being in an inert state, it had not been able to extricate itself from the fork before the waters fell. It was dragged out to the open country by two horses, and was found to measure more than 11m [35 feet 9 inches] in length; on opening it, the

bones of a horse, in a somewhat broken condition, and the flesh in a half digested state, were found within it, the bones of the head being uninjured; from those circumstances we concluded that the Boa had devoured the horse entire.[18]

Size aside, boa constrictors are not aquatic snakes, preferring life in trees - where they are never stranded. On balance, it appears that Gardner met a cousin of the red-tailed boa, who makes up in size for any disappointment occasioned by smaller relatives.

Anacondas

Mystery surrounds the anaconda, from the origin of its popular name to the number of living species and its maximum natural size. Anacondas belong to the genus *Eunectes* ("good swimmer," in Greek), but linguists debate the source of the word *anaconda*. Some trace it to the Sinhalese *henakandaya* ("whip snake"), while others cite the Tamil *anaikondran* ("elephant killer"). Be that as it may, there are no anacondas in the Asian wild, nor any elephants in South America. If anacondas somehow stole their name from Asia's pythons, the date and mechanism of transference remains obscure.[19]

Local names for the anaconda confuse matters further. In Spanish, they are sometimes called *matatoro* ("bull-killer"), while aboriginal names include *boiúba, boiúna, camoodi, guaku, kamudi, kobra detrado, lampalagua, liboya, yakumama,* and *yaurinka.* Several of those names translate as "boa," and are applied impartially by forest-dwelling tribesmen both to anacondas and boa constrictors.[20]

All herpetologists acknowledge two anaconda species, the green anaconda (*Eunectes murinus*) and its smaller relative the yellow anaconda (*E. notaeus*). Two other species, championed by some authors and rejected by others, are Barbour's anaconda (*E. barbouri*) De Schaunensee's anaconda (*E. deschauenseei*), also sometimes called the dark-spotted anaconda. Further confounding the matter are two proposed subspecies of *E. murinus,* listed by some sources as *E. m. gigas* and *E. m. murinus*. Conservative herpetologists suggest that all of the species and subspecies except *E. murinus* and *E. notaeus* comprise minor variants of one or the other, with altered colouration produced by geographic isolation and inbreeding. In fact, the only specimens on record for *E. barbouri* and *E. deschauenseei* were obtained at different times, by different researchers, from Marajo Island at the mouth of the Amazon.[21]

How big do anacondas grow? Again, the record is confused, but all sources attribute greater size to *E. murinus*. Indeed, its yellow cousin is a relative pygmy, with average adult lengths below 10 feet. Thomas Dobson, in his *Encyclopedia of the Arts and Sciences* (1798), describes a 36-foot specimen shot in the Berbice region of Guyana and "sent to the Hague," where it apparently vanished. In 1944, geologist Roberto Lamon led an oil-prospecting party through the jungles of Colombia and shot an anaconda that measured 37 feet 6 inches - but while the group sat down to lunch, before skinning the snake, it revived and escaped. The *Guinness Book of World Records* subsequently deleted that report, while Clifford Pope terms it "probable" and author Chris Mattison considers it suitably "down-to-earth." A 38-footer reportedly killed in Brazil by General Cândido Rondon (1865-1958) - founder of Brazil's Indian Protection Service, for whom the state of Rondônia is named - is likewise unconfirmed.[22]

Other reports are more modest. Vincent Roth, former director of British Guiana's national museum, claimed to have shot a 34-foot anaconda in the 1920s. Dr. Federico Medem of Colombia University sites a specimen 33 feet 8 inches long, killed in the Guaviare River, while naturalist Richard Mole reports a 33-footer found on Trinidad in 1924. American explorer Hyatt Verrill maintained that anacondas never exceed 20 feet, while herpetologist Raymond Ditmars set the limit 12 inches shorter. Disputing those judgments, American herpetologist Thomas Barbour, Brazilian expert Afrânio do Amaral, and José

Cândido de Melo Carvalho of Rio de Janeiro's National Museum all believed that anacondas may exceed 45 feet in length. British naturalist Henry Bates (1825-92) collected 14,000 specimens from the Amazon jungles in 1848-52, including 8,000 previously unknown to science, and mentioned reports of anacondas reaching 42 feet.[23] Supporting evidence for such large specimens is purely anecdotal. The best-known story of a giant anaconda comes from British explorer Percy Fawcett (1867-1925). As prelude to that story, Fawcett wrote, in his memoirs:

> "The manager at Yorongas [in western Brazil, on the Rio Acre] told me he killed an anaconda fifty-eight feet long in the Lower Amazon. I was inclined to look on this as an exaggeration at the time, but later, as I shall tell, we shot one even larger than that."[24] That incident occurred in January 1907, and Fawcett described it as follows:

> We were drifting easily along on the sluggish current not far below the confluence of the Rio Negro when almost under the bow of the *igarité* there appeared a triangular head and several feet of undulating body. It was a giant anaconda. I sprang for my rifle as the creature began to make its way up the bank, and hardly waiting to aim smashed a .44 soft-nosed bullet into its spine, ten feet below the wicked head. At once there was a flurry of foam, and several heavy thumps against the boat's keel, shaking us as though we had run on a snag.

> With great difficulty I persuaded the Indian crew to turn in shorewards. They were so frightened that the whites showed all round their popping eyes, and in the moment of firing I had heard their terrified voices begging me not to shoot lest the monster destroy the boat and kill everyone on board, for not only do these creatures attack boats when injured, but also there is great danger from their mates.

> We stepped ashore and approached the reptile with caution. It was out of action, but shivers ran up and down the body like puffs of wind on a mountain tarn. As far as it was possible to measure, a length of forty-five feet lay out of the water, and seventeen in it, making a total length of sixty-two feet. Its body was not thick for such a colossal length - not more than twelve inches in diameter - but it had probably been long without food....

> Such large specimens as this may not be common, but the trails in the swamps reach a width of six feet and support the statements of Indians and rubber pickers that the anaconda sometimes reaches an incredible size, altogether dwarfing the one shot by me. The Brazilian Boundary Commission told me of one they killed in the Rio Paraguay exceeding *eighty* feet in length! In the Araguaya and Tocantins basins there is a black variety known as the *Dormidera*, or "Sleeper," from the loud snoring noise it makes. It is reputed to reach a huge size, but I never saw one.[25]

While critics in London branded Fawcett a liar, adventurer Algot Lange hunted anacondas of similar size. An opera singer's son who left home at age 26, in 1910, to seek adventure in the wilds of Amazonia, Lange published an account of his exploits in 1912.[26] His description of the region's giant anacondas deserves reproduction in full.

> During my walks in the forest I often came across snakes of considerable length, but never found any difficulty in killing them, as they were sluggish in their movements and seemed to be inoffensive. The rubber-workers, who had no doubt had many encounters with reptiles, told me about large *sucurujus* or boa-constrictors, which had their homes in the river

not many miles from headquarters. They told me that these snakes were in possession of hypnotic powers, but this, like many other assertions, should be taken with a large grain of salt. However, I will relate an incident which occurred while I lived at Floresta [Alta Floresta, in the Brazilian state of Mato Grosso], and in which I have absolute faith, as I had the opportunity of talking to the persons involved in the affair.

Jose Perreira, a rubber-worker, had left headquarters after having delivered his weekly report on the rubber extracted, and was paddling his canoe at a good rate down the stream, expecting to reach his hut before midnight. Arriving at a recess in the banks formed by the confluence of a small creek called Igarape do Inferno, or the Creek of Hell, he thought that he heard the noise of some game, probably a deer or tapir, drinking, and he silently ran his canoe to the shore, where he fastened it to a branch, at the same time holding his rifle in readiness. Finally, as he saw nothing, he returned to the canoe and continued his way down-stream. Hardly more than ten yards from the spot, he stopped again and listened. He heard only the distant howling of a monkey. This he was used to on his nightly trips. No! there was something else! He could not say it was a sound. It was a strange something that called him back to the bank that he had left but a few minutes before. He fastened his canoe again to the same branch and crept up to the same place, feeling very uneasy and uncomfortable, but seeing nothing that could alarm him - nothing that he could draw the bead of his rifle on. Yet, something there was! For the second time he left, without being able to account for the mysterious force that lured him to this gloomy, moon-lit place on the dark, treacherous bank. In setting out in the stream again he decided to fight off the uncanny, unexplainable feeling that had called him back, but scarcely a stone's throw from the bank he had the same desire to return, - a desire that he had never before experienced. He went again, and looked, and meditated over the thing that he did not understand.

He had not drunk *cachassa* that day and was consequently quite sober; he had not had fever for two weeks and was in good health physically as well as mentally; he had never so much indulged in the dissipations of civilisation that his nerves had been affected; he had lived all his life in these surroundings and knew no fear of man or beast. And now, this splendid type of manhood, free and unbound in his thoughts and unprejudiced by superstition, broke down completely and hid his face in his hands, sobbing like a child in a dark room afraid of ghosts. He had been called to this spot three times without knowing the cause, and now, the mysterious force attracting him, as a magnet does a piece of iron, he was unable to move. Helpless as a child he awaited his fate.

Luckily three workers from headquarters happened to pass on their way to their homes, which lay not far above the "Creek of Hell," and when they heard sobbing from the bank they called out.

The hypnotised *seringueiro* managed to state that he had three times been forced, by some strange power, to the spot where he now was, unable to get away, and that he was deadly frightened. The rubber-workers, with rifles cocked, approached in their canoe, fully prepared to meet a jaguar, but when only a few yards from their comrade they saw directly under the root where the man was sitting the head of a monstrous boa-constrictor, its eyes fastened on its prey. Though it was only a few feet from him, he had been unable to see it.

One of the men took good aim and fired, crushing the head of the snake, and breaking the spell, but the intended victim was completely played out and had to lie down in the bottom of the canoe, shivering as if with ague.

The others took pains to measure the length of the snake before leaving. It was 79 *palmas* or 52 feet 8 inches. In circumference it measured 11 *palmas*, corresponding to a diameter of 28 inches. Its mouth, they said, was two *palmas* or sixteen inches, but how they mean this to be understood I do not know.

This event happened while I was living at headquarters. I had a long talk with Perreira, but could not shake his statement, nor that of the three others; nevertheless, I remained a sceptic as to this alleged charming or mesmeric power of the snakes, at least so far as man is concerned.

At that time we were awaiting the arrival of the monthly launch from the town of Remate de Males, and had spent a day weighing rubber at the camp of one of the employees, half a day's journey from headquarters. The rubber-pellets were loaded into our large canoe to take up to Floresta. We spent the evening drinking black coffee and eating some large, sweet pineapples, whereafter we all took a nap lasting until midnight, when we got up to start on our night trip. It had been considered best to travel at night, when it was nice and cool with none of the pestering insects to torture us, and we were soon paddling the heavy canoe at a merry rate, smoking our pipes and singing in the still, dark night. Soon we rounded a point where the mighty trees, covered with orchids and other parasitic plants, sent their branches down to the very water which in its depths was hiding the dreaded water-snakes. The only sound we heard was the weird calling of the night-owl, the "Mother of the Moon" as the Indians call it. Except this and the lapping sound of water, as we sped along, nothing disturbed the tranquility of the night.

I was in the act of lighting another pipe when one of the men cried out: "What's this?"

We all stopped paddling and stared ahead at a large dark object, resting on a moon-lit sand-bar not far from us. Then someone said, "*Sucuruju.*" Few people can comprehend the feeling that creeps into one's heart when this word is pronounced, under such circumstances, in the far-off forest, in the middle of the night. The word means boa-constrictor, but it meant a lot more at this moment. An indescribable feeling of awe seized me. I knew now that I was to face the awful master of the swamps, the great silent monster of the river, of which so much had been said, and which so few ever meet in its lair.

Running the canoe ashore we advanced in single file. I now had a chance to inspect the object. On a soft, muddy sand-bar, half hidden by dead branches, I beheld a somewhat cone-shaped mass about seven feet in height. From the base of this came the neck and head of the snake, flat on the ground, with beady eyes staring at us as we slowly advanced and stopped. The snake was coiled, forming an enormous pile of round, scaly monstrosity, large enough to crush us all to death at once. We had stopped at a distance of about fifteen feet from him, and looked at each other. I felt as if I were spellbound, unable to move a step farther or even to think or act on my own initiative.

The snake still made no move, but in the clear moonlight I could see its body expand and contract in breathing; its yellow eyes seeming to radiate a phosphorescent light. I felt no fear, nor any inclination to retreat, yet I was now facing a beast that few men had ever succeeded in seeing. Thus we stood looking at each other, scarcely moving an eyelid, while the great silent monster looked at us. I slid my right hand down to the holster of my automatic pistol, the 9mm Luger, and slowly removed the safety lock, at the same time staring into the faces of the men. In this manner I was less under the spell of the mesmerism of the snake, and could to some extent think and act. I wheeled around while I still held control of my faculties, and, perceiving a slight movement of the snake's coils, I fired point-blank at the head, letting go the entire chamber of soft-nose bullets. Instantly the other men woke up from their trance and in their turn fired, emptying their Winchesters into the huge head, which by this time was raised to a great height above us, loudly hissing in agony.

Our wild yelling echoed through the deep forest. The snake uncoiled itself and writhing with pain made for the water's edge. By this time we were relieved of the terrible suspense, but we took care to keep at a respectful distance from the struggling reptile and the powerful lashing of its tail, which would have killed a man with one blow.

After half an hour the struggles grew weaker, yet we hesitated to approach even when it seemed quiet and had its head and a portion of its body submerged in the water.

We decided to stay through the night and wait here a day, as I was very anxious to skin the snake and take the trophy home to the States as a souvenir of a night's adventure in this far-off jungle of the Amazon. We went up in the bushes and lit a fire, suspended our hammocks to some tree-trunks, and slept soundly not more than ten yards from the dying leviathan.

We all got up before sunrise, had our coffee in haste, and ran down to see the snake. It was dead, its head practically shot to pieces. We set to work, stretching the huge body out on the sand-bar, and by eight o'clock we had the entire snake flat on the ground, ready to measure and skin.

It was a most astonishing sight, that giant snake lying there full length, while around it gathered six Amazon Indians and the one solitary New Yorker, here in the woods about as far from civilisation as it is possible to get. I proceeded to take measurements and used the span between my thumb and little finger tips as a unit, knowing that this was exactly eight inches.

Beginning at the mouth of the snake, I continued to the end and found that this unit was contained eighty-four times. Thus 84 times 8 divided by 12 gives exactly 56 feet as the total length. In circumference, the unit, the "*palma*," was contained 8 times and a fraction, around the thickest part of the body. From this I derived the diameter 2 feet 1 inch.

These measurements are the result of very careful work. I went from the tail to the nose over again so as to eliminate any error, and then asked the men with me also to take careful measurements in their own manner, which only confirmed the figures given above.

Then we proceeded to skin the snake, which was no easy task under the fierce sun now baking our backs. Great flocks of *urubus*, or vultures, had smelled the carcass and were circling above our heads waiting

for their share of the spoils. Each man had his section to work on, using a wooden club and his machete. The snake had been laid on its belly and it was split open, following the spinal column throughout its length, the ventral part being far too hard and unyielding. About two o'clock in the afternoon we had the work finished and the carcass was thrown into the river, where it was instantly set upon by the vigilant *piranhas* and alligators.

Standing in front of this immense skin I could not withhold my elation.

"Men," I said, "here am I on this the 29th day of July, 1910, standing before a snake-skin the size of which is wonderful. When I return to my people in the United States of America, and tell them that I have seen and killed a boa-constrictor nearly eighteen metres in length, they will laugh and call me a man with a bad tongue." ...

We brought the skin to headquarters, where I prepared it with arsenical soap and boxed it for later shipment to New York. The skin measured, when dried, 54 feet 8 inches, with a width of 5 feet 1 inch.[27]

Explorer Fritz Up de Graff described another huge anaconda in 1923, writing:

"It measured fifty feet for certainty, and probably nearer sixty. This I know from the position in which it lay. Our canoe was a twenty-four footer; the snake's head was ten or twelve feet beyond the bow; its tail was a good four feet beyond the stern; the center of its body was looped up into a huge S whose length was the length of our dugout and whose breadth was a good five feet."[28]

Supporting testimony of a sort comes from Belgium's Marquis de Wavrin (1888-1971), whom Bernard Heuvelmans described as "a careful observer ... [whose] reports are always unromantic and unemotional." The marquis claimed one personal encounter with a 33-foot anaconda, adding that "there are even larger ones, if the natives are to be believed." He once shot a 26-footer, but was dissuaded from fishing it out of the river by native guides who told him "that it was a waste of powder to shoot such a small snake and a waste of time to pick it up." They added: "On the Rio Guaviare, during floods, chiefly in certain lagoons in the neighbourhood, and even near the confluence of this stream, we often see snakes which are more than double the size of the one you have just shot. They are often thicker than our canoe."[29]

In January 1947, after Chavante Indians killed several government officials, Brazil's Service for the Protection of Indians organized a 20-man expedition to restore peaceful relations with tribesmen along the Rio Araguaia. Two Frenchmen, artist Serge Bonacase and explorer Raymond Maufrais, joined the team, which traversed the river between February and July. In late March, while camped on an island between the Rio Cristalino and Rio Manso, Bonacase joined a hunting party and killed a huge anaconda. The hunters measured their prey with a three-foot piece of string and found it to be "nearly 23 metres [74 feet 9 inches] long." Describing the event to Bernard Heuvelmans, Bonacase described the snake's head by holding his arms out with his fingers interlaced, creating a triangle with 25-inch sides and an 18-inch base.[30] When Heuvelmans asked why no part of the snake was preserved, Bonacase replied:

> First of all, none of us seemed to realise that there was anything exceptional about our prize. There were no zoologists among us. The Brazilian officials who had spent much of their lives in this country did not seem to be particularly surprised. As for me, I had heard so many tales of giant snakes that I supposed the whole of the Amazon was crawling with monsters of this size.
>
> Of course, we should have liked to take the snake's skin back, but we had neither the time to skin the beast nor to prepare its hide. We were also very tired, and in that country you must never let yourself linger by the way or encumber yourself with extra luggage. Think what a piece of skin more than 20m [65 feet] by 1.40 m [4 feet 6 inches] wide would have weighed! We should have had out work cut out to carry it - or for that matter the head. Besides, who would have been crazy enough to lug a piece of rotting meat on his back in that heat through country infested with insects? [31]

A report of a somewhat smaller anaconda comes second-hand from David Atlee Phillips (1922-88), a 25-year veteran of the Central Intelligence Agency who heard the tale from two fellow spies. Those CIA agents, in turn, claimed direct participation in the reptile's capture, sometime in the early 1950s.

As Phillips related the tale, a Bolivian cattle rancher approached a CIA agent identified only as "Lee," complaining of a snake more than 33 feet long which drowned steers in a river and "had eaten, at the very least, ten Indians." Lee and a fellow agent planned to catch what "was certainly the largest snake in existence," using a long cotton-picking sack with zippers at each end. After months of debate, they stormed the snake's cave, flushed it out with tear gas, and bagged it - but the bag ripped at its seams and they were forced to shoot their prize. It measured 34 feet 3 inches, and Lee displayed its skin to Phillips.[32]

Some authors suggest that if snakes of such size ever lived in Amazonia, they must now be extinct. Contrary evidence emerged in 2007, when Josh Gates and company pursued legends of giant anacondas on behalf of the Sci-Fi Channel's program *Destination Truth*. One tribesman interviewed on camera described an attack by a snake 45 feet 6 inches long, which snatched him from a boat and tried to drown him before it was killed. An even larger serpent, estimated at 65 feet, had raided a riverside village and devoured several dogs. A witness measured the anaconda's head, as Bonacase had done, by forming a triangle with his outstretched arms.[33]

We can imagine the result of meeting such huge serpents in the wild, particularly if the humans are inadequately armed. Three reports from the 1950s describe victims slain - and in one case, devoured - by anacondas of "ordinary" size. Rolf Blomberg, writing for *Natural History* in 1956, detailed two cases from Ecuador.

The first victim, a 13-year-old boy, was snatched while swimming with friends at the mouth of the Rio Yasuní, a tributary of the larger Rio Napo. A protracted search led hunters to the snake, which had apparently swallowed, then regurgitated its victim before it was shot. A second, adult victim was killed by an anaconda while swimming in the Rio Napo itself. Author Kurt Severin, writing for *True* magazine in June 1958, supplied the "authenticated" case of a man seized and drowned while watering his cattle.[34]

Anacondas ranging from 40 to 75 feet may strain credulity for herpetologists, and while none has been formally documented, it appears that they are not the largest snakes in South America. *Much* larger specimens have been reported over the past century - so large, in fact, that most cryptozoologists regard them as members of a separate unknown species.

Sucuriju Gigante

As with the anaconda, we begin our survey of these giant serpents mired down in a Babel of popular names. The most common name found in print is *sucuriju* (or *sucuruju*) *gigante,* translated as "giant boa." Other names employed for the same creature(s) include *cobra grande* in Brazil, *vulpangue* in Chile, and *sachamama* ("mother of the Earth") in Peru.[35] It seems that all describe the same titanic reptiles, which easily dwarf the largest reported anacondas.

Lorenz Hagenbeck, son of famed German animal collector Carl Hagenbeck, heard the first known eyewitness accounts of the *sucuriju gigante* from two priests serving in Brazil, Father Victor Heinz and a Father Protesius Frickel. Father Heinz allegedly saw giant snakes on at least three occasions between 1922 and 1929; he also regaled Hagenbeck with the tales of two independent witnesses. Father Frickel claimed only one sighting, which Hagenbeck endorsed without reservations.[36]

Father Heinz's first encounter occurred on 22 May 1922, while he was traveling upriver by canoe, from Óbidos on the Amazon. Recent floods had filled the river with debris, some of which was alive. As Heinz described the incident:

> [S]uddenly I noticed something surprising in midstream. I distinctly recognized a giant water-snake at a distance of some 30 yards. To distinguish it from the *sucuriju,* the natives who accompanied me named the reptile, because of its enormous size, *sucuriju gigante* (giant boa).

> Coiled up in two rings the monster drifted quietly and gently downstream. My quaking crew had stopped paddling. Thunderstruck, we all stared at the frightful beast. I reckoned that its body was as thick as an oil-drum and that its visible length was some 80 feet. When we were far enough away and my boatmen dared to speak again they said that the monster could have crushed us like a box of matches if it had not previously consumed several large capybaras.[37]

> Coincidentally, Heinz claimed that a similar serpent had lately been killed at Lago Grande de Salea, a day's march south of Óbidos. Tribesmen surprised the reptile as it was swallowing a capybara, and found four more of the same when they opened its stomach. Capybaras are Earth's largest living rodents, growing to 4 feet 4 inches long and weighing up to 140 pounds at maturity. Nearby, tribesmen found two large piles of excrement containing animal hair and a bone from the leg of an ox, including its hoof.[38]

Father Heinz met his next *sucuriju gigante* on 29 October 1929, while traveling downriver from Alemquer. He departed at 7 p.m. and encountered the monster five hours later. As Heinz described the event:

> At about midnight, we found ourselves above the mouth of the Piaba when my crew, seized by sudden fear, began to row hard towards the shore.

> "What is it?" I cried, sitting up.
> "There is a big animal," they muttered, very excited.

At the same moment I heard the water move as if a steamboat had passed. I immediately noticed several m[eters] above the surface of the water two bluish-green lights like the navigation lights on the bridge of a river-steamer and shouted:

"No, look, it's the steamer! Row to the side so that it doesn't upset us."

"*Que vapor que nada,*" they replied. "*Una cobra grande!*"

Petrified, we all watched the monster approach; it avoided us and recrossed the river in less than a minute, a crossing which would have taken us in calm water ten to fifteen times as long. On the safety of dry land we took courage and shouted to attract attention to the snake. At this very moment, a human figure began to wave an oil-lamp on the other shore, thinking no doubt, that someone was in danger. Almost at once the snake rose on the surface and we were able to appreciate clearly the difference between the light of the lamp and the phosphorescent light of the monster's eyes. Later, on my return, the inhabitants of this place assured me that above the mouth of the Piaba there dwelt a *sucuriju gigante*.[39]

Energized by his personal encounters with the giant snakes, Father Heinz began collecting tales from other witnesses, which he relayed to Hagenbeck. The first was that of Reymondo Zima, a Portuguese merchant who spent nine years at Faro, on the Rio Jamunda (or Nhamundá). Zima declared:

On 6 July 1930 I was going up the Jamunda in company with my wife and the boy who looks after my motor-boat. Night was falling when we saw a light on the right bank. In the belief that it was a house I was looking for I steered towards the light and switched on my searchlight. But then suddenly we noticed that the light was charging towards us at an incredible speed. A huge wave lifted the bow of the boat and almost made it capsize. My wife screamed in terror.

At the same moment we made out the shape of a giant snake rising out of the water and performing a St. Vitus's dance around the boat. After which the monster crossed this tributary of the Amazon about half a km wide at fabulous speed, leaving a huge wake, larger than any of the steamboats make at full speed. The waves hit our 13m boat with such force that at every moment we were in danger of capsizing. I opened my motor flat out and made for dry land.

Owing to the understandable excitement at the time it was not possible for me to reckon the monster's length. I presume that as a result of a wound the animal lost one eye, since I saw only one light. I think the giant snake must have mistaken out searchlight for the eye of one of his fellow-snakes.[40]

The final incident related by Father Heinz occurred 12 weeks later, on 27 September 1930. The witness in that case, João Penha, was clearing a bank of the Rio Iguarapé, near Lake Maruricana, for the benefit of spawning turtles. Heinz described what happened next:

At a certain moment, behind one of those floating barriers made of plants, tree-trunks and tangled branches, against which steamers of 500 tons often have to battle to force a passage, he saw two green lights.

Penha thought at first that it was some fisherman who was looking for eggs. But suddenly the whole barrier shook for 100m, He had to retreat hurriedly for a foaming wave 2m high struck the bank. Then he called his two sons, and all three of them saw a snake rising out of the water pushing the barrier in front of it for a distance of some 30m until the narrow arm of water was finally freed of it. During all this time they could observe at leisure its phosphorescent eyes and the huge teeth of its lower jaw.[41]

Two or three years after that encounter (reports differ), a *sucuriju gigante* was killed by members of the Brazilian Boundary Commission on the Rio Negro, somewhere near the Venezuelan frontier. Witnesses to the encounter, who dispatched the snake with machine-gun fire, described the reptile as 97 feet 6 inches long and 24 inches in diameter. Four men were unable to lift the snake's head, prompting an estimate of its total weight at two tons. A blurry photo of the snake was published by an unnamed Brazilian newspaper in 1933, and while the original has long been lost, it survives on the Internet, sadly providing no objects useful in judging scale.[42]

Two sightings of the *sucuriju gigante* emerged in 1948. The first involved witness Paul Tarvalho, a former pupil of Father Heinz who glimpsed a giant snake near the same place where Heinz logged his sighting in October 1929. According to Tarvalho, he first saw the serpent emerge from the water, then it briefly pursued his boat, before he escaped at top speed. He estimated the creature's length at 163 feet.[43]

The second incident from 1948 also produced another startling photograph. Soldiers found the snake in question living in the ruins of Fort Tabatinga, on the Rio Oiapoque, which forms the boundary between Brazil and French Guiana. Reportedly, it took 500 bullets to kill the monster, which measured either 114 or 131 feet (reports differ). To eliminate the giant carcass, soldiers pushed it into the river, then snapped a photo as it drifted away. That photo was published in the newspaper *A Provincia do Pará* on 28 April 1949. An alternative version of the tale names the river in question as the Rio Abuna, in the Brazilian state of Acre - located some 1,500 miles west of the Rio Oiapoque.[44]

Author Mike Dash confuses matters even more, with a summary of the Brazilian photographic evidence, published in 2000. Specifically, he writes:

> Four photographs, published in provincial Brazilian newspapers, do purport to show giant boa constrictors [*sic*] in the Amazon jungle. One is a head-on shot of what was said to be a 130-foot-long, five-ton specimen, killed near Manaos; another, a snapshot of what is said to be a 100-foot-long corpse drifting downriver at the mouth of the Amazon; a third depicts a giant of 98 feet, killed on the banks of the River Negro; and the last portrays a snake, purportedly 115 feet long, which was dispatched in a ruined fort by 500 bullets from a machine gun. All, however, are of limited value as each lacks a convincing indication of scale.[45]

Dash produces none of the photos in question, nor do any of his sources listed for the passage quoted here refer to four existing photographs. On balance, it appears that Dash became confused during production of his manuscript and multiplied two photos into four. The first and third apparently refer to the 1933 photo, while numbers two and four describe the 1948 incident.

Reports of giant snakes in Amazonia did not end in 1948, with the slaying of Fort Tabatinga's scaly invader. Indeed, a report from Rio de Janeiro, published in February 1969, announced that Italian naturalist Bruno Falcci was embarking on a search for a serpent "more than 100 feet long and at least a yard wide." Falcci claimed to have seen the monster sleeping beside aptly-named Surprise Lake, 180 miles southwest of Porto Velho, capital of the Brazilian state of Rondônia, abutting Bolivia. Lacking a means to catch the huge reptile, Falcci snapped a photo (never published), then reported his encounter to a priest called Father Bendorraite, serving at a local mission for the native Paccas Novos tribe. The snake was still asleep when Falcci returned with a group of tribesmen, but it woke and fled into the lake while Falcci was persuading the Indians not to kill it.[46]

Falcci had no means of measuring the reptile, but he estimated its 100-plus-foot length from deep impressions left by its body on the lake's shore. He then flew home to Italy, vowing that he would return to capture the serpent alive, but local natives planned to kill it, first. One of the hunters, a Bolivian identified only as "Ramon," blamed the snake for devouring his cousin Pablo, along with Pablo's wife and four children. The family had vanished without explanation in 1966, and Ramon jumped to conclusions three years later, upon hearing Falcci's report. "That's what did away with Pablo, his wife and kids," Ramon told reporters. No follow-ups were published to the early news from 1969, and we may assume that the snake was never found.[47]

British travel author Jeremy Wade went searching for the *cobra grande* in 1995-96, and reported his findings in *Fortean Times,* in May 1997. He began the journey "not expecting much," but quickly found that tales of giant snakes, "many first-hand, confounded expectations."[48] The stories he collected include:

- A sighting logged by "down-to-earth" witness Amarilho Vincente de Oliveira on a tributary of the Rio Purus, in the Brazilian state of Acre (abutting Peru and Bolivia). Wade reports that the event occurred "20 years ago," presumably meaning 1977. The witness described a giant snake whose head "had horns like the roots of a tree, and we could see these greenish eyes as well."[49]
- Witness Dorgival Sabino saw a snake "much bigger than normal" - 65 feet long and 3 feet in diameter - in the Rio Negro, near Manaus. As with Oliveira's specimen, "its head was like some kind of dinosaur, with - I don't know whether they were teeth or horns, just that it was grotesque."[50]
- Unnamed fishermen on the Rio Purus netted a giant snake by accident, then wounded it with a dynamite charge before letting it drift downstream. Weeks later, one witness found a reptilian skull more than 20 inches long, beached with a 24-inch rib on the river's bank.[51]
- A 76-year-old witness regaled Wade with a story from his youth, when he had seen a three-foot rib cut from a snake so large that it could not be dragged ashore, once it was killed. Wade calculated that the rib's owner would have been 24-26 inches in diameter - 50 percent wider than the 74-footer reported by Serge Bonacase in 1947.[52]
- In June 1995, Brazilian botanist Grace Rebelos dos Santos was camped along the Amazon, at a point where local fishermen had netted, then lost, some unseen massive object in the afternoon. That night, Rebelos saw two lights "like torches," 12 inches apart, appear in mid-river and move toward shore, then submerge. She told Wade, "I'm not going to say it was a *cobra grande,* but I remember clearly how blue the lights were, which I thought very strange." Five months later, on 23 December, Rebelos saw a 20-inch-high "waterspout" erupt from the river, above a large dark shape that quickly sank from view.[53]
- Finally, Wade relates two vague stories of "canoes being pulled under and floating up in pieces." He identifies no witnesses, but describes one ex-fisherman who survived an attack with two broken ribs, plus sundry cuts and bruises.[54]

Wade does his best to debunk the *cobra grande,* including an undated incident wherein soldiers were summoned to kill a 60-foot snake at an unnamed lake, "before it got out and ate the village." The "snake" proved to be a tree branch half-buried in mud, which, when pressed on one end, caused the ground to stir 30 feet away. Wade's suspects in canoe-tippings include dolphins and bull sharks. As for horned snakes, he suggests "an optical illusion, arising from big anacondas with very long teeth," or those who swallow deer feet-first and leave antlers protruding from their mouths.[55] However, we shall meet horned giant snakes again in North America (see Chapter 7), where no species large enough to eat a deer officially exist.

Six months after Wade's article hit news-stands, *Fortean Times* carried another startling giant-snake report. On 19 August 1997, a jet-black snake 130 feet long allegedly raided the village of Nuevo Tacna, near the Rio Napo (an Amazon tributary in northern Peru). Jorge Samuel Chávez Sibina, mayor of the Municipalidad Provincial de Maynas (170 miles northeast of Lima), flew over Nueva Tacna with radio journalist Carlos Villareal, and declared, "There were five witnesses present and the rest of the 300 villagers felt the effect of this thing as it dragged itself along and dived into the River Napo. Something strange has taken place. There really is something to the villagers' stories."[56]

But *what?* Skeptics insist that residents of Nueva Tacna are so ignorant and excitable that they mistook an earthquake for a giant black serpent slithering past their homes. That theory fails to account for the monster's track, measured at 1,600 feet long and 30 feet wide, or the several boats that it sank in the Rio Napo. Author Karl Shuker's suggestion that the animal - known locally as *sachamama* - might be some kind of giant unknown mollusk, is even more surprising than the story of a giant snake at large.[57]

Josh Gates and his fellow travelers from *Destination Truth* revealed another recent story of the *sucuriju gigante* in 1997. That eyewitness report, collected from tribesmen along the Amazon, described an undated close encounter with a serpent 98-130 feet long.[58]

The *Miñocao*

Another giant snake reported from Amazonia may not be a reptile at all - but since it remains unidentified more than a century after its existence was first reported, it deserves inclusion here.

German biologist Johann "Fritz" Müller (1821-97) emigrated to Brazil at age 31 and spent the rest of his life cataloging its natural wonders. He was the first European to report tales of the *miñocao* (or *minhocão*, in Portuguese), described by native witnesses as a burrowing animal 160 feet long and 16 feet wide, whose tunnels uprooted trees and diverted rivers. Müller wrote a report on the *miñocao,* declaring that its name described a "giant earthworm."[59]

The first case cited in Müller's report dated from 1849, when farmer João de Deos found his land in the southern Brazilian state of Paraná churned up by some creature that burrowed by night. The strange tracks, still visible three years later, prompted landowner Lebino José dos Santos to speculate that two animals, each three to six feet wide, had dug the furrows. Dos Santos claimed that other local residents saw the beast itself. One said it was "as big as a house"; the other described a black worm-like creature "no longer than a lasso" - that is, 81 feet long - with two "moveable horns" on its head. Yet another witness told Lebino of a *miñocao* that had died while trapped in a rocky crevice near Arapehy, Uruguay. Its skin was thicker than a pine tree's bark, covered with scales like those of an armadillo.[60]

In the 1860s, a rather different *miñocao* appeared to witness Francisco de Amaral Varella near Lage, on the Rio Caveiras, in the southern Brazilian state of Santa Catarina (immediately south of Paraná). Varella described the creature as three feet in diameter, but "not very long." It had a pig-like snout, but no visible legs. It left three-foot-wide trenches before plunging underground, and similar trails appeared on the far side of Lage, weeks later. Meanwhile, a German immigrant surveyor found *miñocao* trenches too wide for him to step across, while plotting the course of a road from Itajahy into Santa Catarina's highlands, along the Rio Marombas.[61]

Around the same time, Antonio José Bosco returned to his rural home on the Rio dos Cachorros, six miles from Curitibanos in Santa Catarina, and found the nearest road demolished by a trench nine feet wide and some 2,600 feet long, terminating in a swamp. Tremors and rumbling noises from the marsh, clearly audible on quiet nights, convinced locals that the *miñocao* was still in residence.[62]

In 1863 burrowing creatures known to locals as *sierpe* ("serpents") wreaked havoc on farms around La Cuchilla, in Santa Catarina. They toppled fruit trees, cast up boulders, and left caved-in furrows behind. Witness Paulino Montenegro described the trenches as four feet deep and five feet wide.[63] Fritz Müller concluded:

> From all these stories it would appear conclusive that in the high district where the Uruguay and the Paraná have their sources, excavations and long trenches are met with, which are undoubtedly the work of some living animal. Generally, if not always, they appear after continued rainy weather, and seem to start from marshes or river-beds, and to enter them again. The accounts as to the size and appearance of the creature are very uncertain. It might be suspected to be a giant fish allied to *Lepidosiren* and *Ceratodus* [species of lungfish]; the "swine's snout" would show some resemblance to *Ceratodus,* while the horns on the body rather point to the front limbs of *Lepidosiren,* if these particulars can be depended upon. In any case it would be worth while to make further investigations about the Minhocăo, and, if possible, to capture it for a zoological garden![64]

Decades later, Percy Fawcett described a similar creature. He wrote:

> In the Paraguay River there is a freshwater shark, huge but toothless, said to attack men and swallow them if it gets a chance. They talk here of another river monster - fish or beaver - which can in a single night tear out a huge section of river bank. The Indians report the tracks of some gigantic animal in the swamps bordering the river, but allege that it has never been seen. The shark exists beyond doubt; as for the other monsters - well, there are queer things yet to be disclosed in this continent of mystery, and if strange, unclassified insects, reptiles and small mammals can exist there, mightn't there be a few giant monsters, remnants of an extinct species, still living out their lives in the security of the vast unexplored swamp areas? In the Madidi, in Bolivia, enormous tracks have been found, and Indians there talk of a huge creature descried at times half submerged in the swamps.[65]

Bernard Heuvelmans concluded that the *miñocao* was not a snake, nor any reptile at all, but "some kind of giant armadillo." Specifically, he settled on a relict *Glyptodon,* a Pleistocene relative of the armadillos and anteaters, presumed extinct for some 10,000 years. Fossil remains demonstrate that the *Glyptodon's* shell exceeded six feet in length, while the creature's overall length rivaled that of a small automobile. Modern armadillos are burrowers, and while they can also swim, no evidence of any aquatic or amphibious species presently exists.[66]

Giants Astray

Nicaragua lies in Central America, technically beyond the scope of this chapter, and - at least officially -

it has no anacondas. Its largest known snakes are the boa constrictor and the venomous bushmaster (*Lachesis muta*), with a record length of 11 feet 10 inches.[67] Still, we have one story of a giant snake from Nicaragua, reported by French author François Poli in 1958, which deserves mention here.

Poli heard the tale from a German named Brennecker, who spent 20 years in Nicaragua and recalled his meeting with a giant reptile near that country's border with Honduras. Poli writes, quoting Brennecker:

> "I was driving a jeep along a sort of natural track winding between two lines of trees when I saw, about fifty yards ahead, a huge fallen tree-trunk which barred the way. I told the boy who was with me to go and find some way of shifting it. He came back at a run. It wasn't a tree-trunk at all, but a snake. It stirred and began moving slowly toward us...."

There was a silence, and then [Brennecker] added:

> "I've seen the most incredible snakes in this country during the past twenty years - and I can assure you that I know how to handle a gun. But that day I left the revolver where it was; I just stepped on the gas and drove off."[68]

A wise decision, all things considered, but it leaves the Nicaraguan mystery - like those of Amazonia and points south - forever unsolved.

The Boas

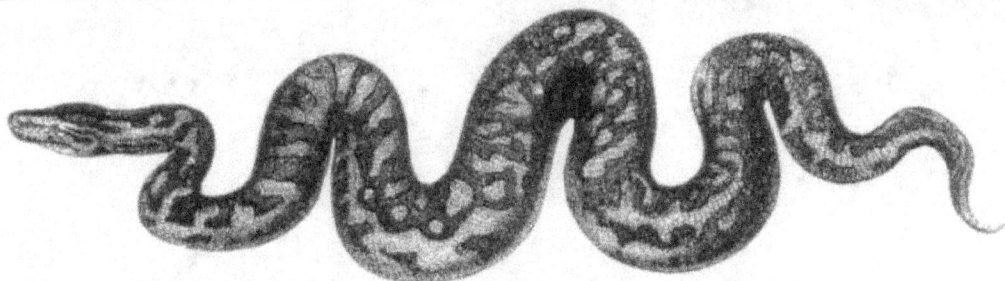

Cerastes ex Libya Aldrou.

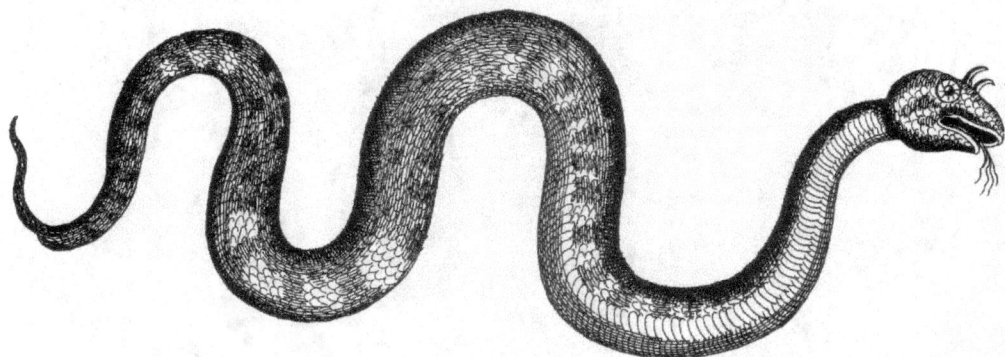

TOP: Vintage photograph of a large reticulated python about to be force-fed
MIDDLE: An early engraving of a boa
BOTTOM: `Cerastes` - a horned serpent depicted in Charles Owen's 1742 book *An Essay towards a Natural History of Serpents - in two parts*. It is probably a horned viper.

A 19th Century engraving of a boa constrictor. The monkeys in the background are capuchins

An early engraving of a python attacking a man

A giant snake photographed during the war of the Katanga Secession in the Congo

FIG. 54.—THE PA SNAKE. (*From the Shan Hai King.*)

The Pa snake—a mythical Chinese reptile said to be ablt to devour whole elephants

A French Foreign Legionnaire fighting a python

Ibiboboca .

Another snake from Owen's book - the Ibiboboca (an early depiction of the giant anaconda)

TOP: An African rock python
BOTTOM: A stylised depiction of an Indian snake

Serpens   Iudicus

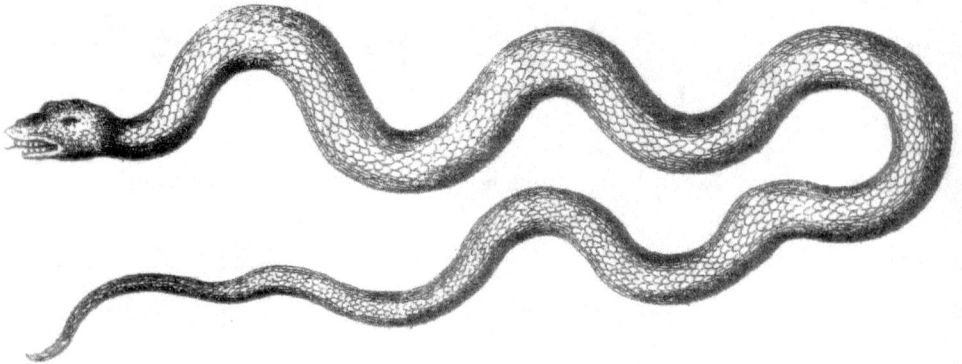

just as I raised the piece to take aim at the monsters head, it had reached the flowing train of Olympia's robe and with a dreadful hiss tore her down from her husbands arms.

London. Pub. April 1. 1803 by T. L...   *Page 26*

John Browne's *ibibaboka* from his book *Affecting Narrative*
An early artistic demonisation of the anaconda

A realistic 19th Century depiction of an anaconda

The Skinning of the *Aboma Snake*, shot by Capt. Stedman' William Blake
Great Britain 1796.

TOP: Another depiction of the incident shown on the previous page
BELOW: Supposed *sucuriju gigante* photograph

Engraving of python

TOP: Alleged 45m giant anaconda shot Guapore territory, Brazil in the 1940s
BOTTOM: North American giant snake chasing a man

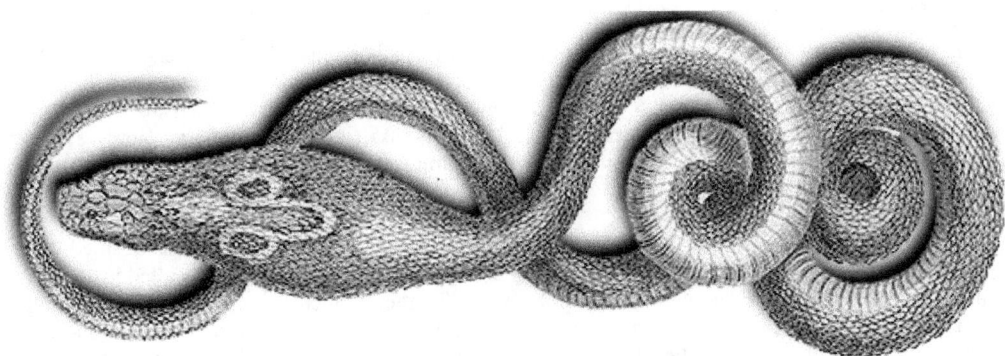

Chapter 7: North America

A measure of confusion surrounds North American ophiology, extending to the number of snake species found in Canada, the U.S.A., and Mexico. A "complete list" of 127 species, posted on the Internet by the Slater Museum of Natural History (University of Puget Sound), omits many recognized subspecies, nor is any consensus derived from the standard field guides: Conant and Collins list 150 species for eastern/central North America (excluding Mexico), while Stebbins names 210 for western North America and Baja California (again, excluding Mexico proper). Naturally, the ranges of some species listed in those field guides overlap. At least 27 snake species are known to inhabit Mexico.[1]

None of the "big six" snakes are North American natives, and while exotic pythons have established breeding colonies in southern Florida, no true giants should be found outside of zoos or private collections. The largest recognized native species include the bullsnake (*Pituophis catenifer*), at 8 feet 9 inches; the indigo snake (*Drymarchon corais*), 8 feet 7.5 inches; the coachwhip (*Masticophis flagellum*), 8 feet 6 inches; the eastern ratsnake (*Pituophis alleghaniensis*), 8 feet 5 inches; and the eastern diamondback rattlesnake (*Crotalus adamanteus*), 8 feet 3 inches. Three other species - the western diamondback rattler, the pine snake, and the western ratsnake - exceed 7 feet. Six others - the common king snake, mud snake, racer, Florida green water snake, eastern cottonmouth, and timber rattlesnake - commonly exceed 6 feet in length.[2]

No giants there. And yet, from prehistoric times to the 21st century, reports of larger snakes, often beyond the length acknowledged for the largest python or anaconda, have been logged throughout the United States and Canada. Today it is common to blame all such sightings on escaped exotic pets, but what should we make of giant-snake traditions that predate the arrival of European settlers?

Horned Serpents

We have examined various reports of large horned snakes from Amazonia and Asia; now we find similar (if not identical) creatures enshrined in the legends of Native Americans. Worldwide, at least a dozen species of snakes - all vipers - sport small horns on their snouts or above their eyes, but none approach giant size, and North America's only horned species (the sidewinders, *Cerastes cerastes* sp.) do not officially exceed 33 inches.[3]

Tales of great horned serpents, dwelling both on land and in various lakes or rivers, pervade Native American mythology. Diverse names for these huge reptiles included *Doonongaes* (among the Seneca), *Kolowisi* (Zuni), *Misikinipik* (Cree), *Stvkwvnaya* (Seminole), *Tijaiha* (Huron), and *To kas* (Klamath). The attitude of horned serpents toward humankind varied from case to case, some being purely evil, while others proved ambivalent and fickle. Still, all native cultures stressed the need to give such animals a wide berth whenever possible.[4]

The Seminole tradition of *Stvkwvnaya,* the 20-foot "tie snake" whose single horn was deemed to be an aphrodisiac, predates Florida's python invasion by at least three centuries. It was recalled in 1902, when hunter Buster Ferrel killed "an enormous reptile, more like the mythical dragon than a land serpent," in the Everglades. Ferrel pegged the snake's length at 20-30 feet, and while vultures devoured most of the carcass, he allegedly retained the creature's head, described as "a frightful looking object, fully ten inches from jaw to jaw, with ugly, razorlike teeth."[5]

A larger serpent, this one sporting proper horns, menaced Indiana farmer Jacob Rishel near Fort Wayne, in August 1879. After surprising Rishel in a grassy field, the beast chased him for 100 yards, until he reached a fence where a scythe and a reaping machine stood waiting. Rishel grabbed the scythe and severed one of the snake's horns (or "tentacles"), then leapt back as the striking monster impaled itself on the reaper. Rishel's next blow beheaded the snake, whereupon the farmer collapsed in a faint. Neighbors found him unconscious, roused him, and helped Rishel measure the snake, which was 34 feet 3 inches long and "about as thick as a barrel." Its horns measured 42 inches long and were 3 inches wide at their base. The *Fort Wayne Sentinel* reported that the snake was skinned, its hide sent "to Chicago" for exhibition, but no more was heard of the trophy.[6]

While some researchers view Rishel's story as a 19th-century newspaper hoax, it produced an eerie echo seven years later, in the Fort Wayne *Weekly Gazette.* According to that story from June 1886, local farmer Henry Vegalcus was pursued for 110 yards by "a monster snake," before he found a scythe hanging from a fence and used it to kill the reptile. The serpent had no horns, but measured a respectable 17 feet long and 2 feet in diameter. Its dorsal scales were black, its belly "a sort of red."[7]

Did either Hoosier snake exist, in fact? Was the 1886 report simply scaled-down plagiarism of the Rishel story, or did 19th-century Indiana breed large unknown snakes with a penchant for stalking farmers? Since neither specimen was preserved or photographed, such questions must remain unanswered.

Eleven years after the second Fort Wayne incident, in May 1897, "an enormous reptile" at least 50 feet long appeared in Phillips County, Kansas, on the Nebraska border. Witnesses reported that it had "the sinuous body of a snake, but its tongue is not forked, and on its head are two short horns. Its colour is green, with dirty white spots." The serpent was blamed for devouring 40 young pigs on one farm, 60 chickens on another, and for flogging a horse to death with its tail. Bullets bounced off its hide, while its three-foot-long tongue lashed the air and it whined "like a puppy crying for its mother."[8]

A quarter century later, in June 1931, hunting parties organized to stalk a "monster snake" around Luray, Kansas, 60-odd miles to the southwest, in Russell County. A cartoon of "Luray's famous snake," published in May 1938, depicts a serpent large enough to swallow a man, with a single horn protruding from its forehead. As late as 2001, Luray's Friendship Day parade featured a 46-foot replica of the snake - this, despite a statement from college student Leslie Doane, published in October 1936, confessing that he and an unnamed companion had manufactured the 1931 snake from "canvas, excelsior and paint," displaying it at various points in Russell County and adjoining Osborne County (which lies between Russell and Phillips County).[9]

Where the Boas Are

It may be no coincidence that despite their harsh winters, the Midwestern states of Illinois, Indiana and Ohio have produced so many giant snake sightings from Colonial times to the present day. Chad Arment's epic catalog of North American "boss snake" reports lists 7 Illinois counties, 18 in Indiana, and 32 in Ohio with giant snake reports on file between 1793 and 1961. Author Mark Hall explains that prevalence of serpents, at least hypothetically, by noting that vast swamps once covered much of the Midwest.[10]

Specifically, the Great Black Swamp sprawled across northwestern Ohio and northeastern Indiana, including the watersheds of the Auglaize, Blanchard, Maumee, Ottawa, Portage, Sandusky and Tiffin Rivers. It stretched roughly from present-day Sandusky and Toledo, Ohio, to New Haven, Indiana. Parts of the Great Black Swamp remain today as the Ottawa National Wildlife Refuge. Indiana's Limberlost Swamp covered 13,000 acres of Adams and Jay Counties, drawing its name from the 19th-century disappearance of "Limber Jim" Miller. A larger expanse of wetlands, dubbed the Prairie Pothole Region, presently supports more than 50 percent of North America's migratory waterfowl. Its vast expanse includes northern Iowa, eastern Minnesota, roughly half of North and South Dakota, a strip of northern Montana, southwestern Manitoba, the southern half of Saskatchewan, and more than one-third of Alberta.[11]

Pennsylvania boasts giant-snake reports from 27 of its 67 counties since 1820, but many sightings emerge from Adams County, site of the Gettysburg Civil War battlefield. Big Round Top, a hill two miles south of Gettysburg, has produced frequent sightings, while others emerge from the Devil's Den, on Houck's Ridge near Little Round Top. In fact, while published mentions of Devil's Den all date from the post-Civil War period, historian John Bachelder's 1873 guidebook to Gettysburg claimed that Devil's Den "was a name given to the locality before the battle." Local author Emanuel Bushman, writing in 1884, described a "monster snake" inhabiting Devil's Den. Three decades later, in 1913, Salome Stewart claimed that an elusive reptile called "The Devil" lent its name to Devil's Den.[12]

North America's monster-snake reports begin in Colonial times. The first reports involve huge rattlesnakes (discussed in more detail below), but sadly no specific locations are known for the slaying of rattlers measuring 17 feet (1712), 18 feet (1750), and 22 feet (1753).

We know a great deal more about the "American anaconda" killed by hunters west of Fort Recovery - in Mercer County, Ohio - on 27 May 1793. As described two years later, in Rutland, Vermont's *Rural Magazine,* the snake measured 26 feet 7.5 inches "and was thick in proportion."[13] Furthermore:

> His head was green, with a large black spot in the middle; round the jaws, which were very flat, but extremely broad were great streaks, and his eyes were monstrously large, very bright and terrible. His sides were formed of streaks of bright red, green, white, purple, and pale blue, and more beautiful than can be well imagined. Down his back ran a broad stroke of olive green, twisted and waved at the edges, beside which was a narrow one of flesh colour; and on the outside of that a very broad one of bright yellow, waved and curled in various inflections.[14]

The hunters cut open their prize, revealing in its stomach "a panther, several squirrels of different species, birds, insects, and snakes of an inferior kind, all of which had been swallowed whole, and not a bone broken." The reptile's hide was sent off to "the Philadelphia museum," and thereupon vanished forever.[15]

A quarter-century elapsed before the next reported incident, from Alabama, where an unseen serpent left huge tracks around Greenville in 1818. An 11-footer, seen in Pennsylvania two years later, was dwarfed by those seen in Québec and Maryland, in 1826 and 1829, respectively. Farmers blamed the Canadian 30-footer for eating "a horse, six or seven neat cattle, and some sheep." Maryland's monster exceeded 30 feet, with a body "as large as a common saw log."[16]

The 1830s produced two sightings. One, from Kentucky in 1830, involved a snake 18-20 feet long and "as thick as an ordinary stove pipe." The other, from Pennsylvania in 1833, described a black snake "turning gray with age," 15-20 feet long.[17]

Sightings doubled in the 1840s, with four reports logged from three states. Massachusetts scored first, in July 1843, with the slaying of a snake 14 feet 3 inches long and 1 foot 10.5 inches in circumference. Ohio came next, in January 1846, with claims of a snake "fully 30 feet" long. Vermonters saw a 12-foot serpent as thick as the proverbial stove pipe, in August 1848. One year later, Ohio produced more sightings of a snake 30-35 feet long.[18]

Reports nearly doubled again in the 1850s, with seven sightings from as many states. New Yorkers beat the bushes for a 12-foot reptile and its 3-foot offspring in July 1853, while September 1853 brought a 20-footer to Maryland's Allegany Valley. A 14-foot reptile frightened West Virginia residents in 1854, while some reporters dismissed sightings of a giant snake at Cairo, Illinois, that same year as the work of hoaxers. Alabama's serpent returned in 1855, leaving tracks 18 inches wide around Greenville. Two years later, hunters killed a 20-foot rattler in Kentucky. Pennsylvania finished off the decade, with an 1859 report of a snake 10-15 feet long and "as thick as a man's leg."[19]

America's civil war did not prevent witnesses in eight states from logging 15 giant-snake sightings in the 1860s. A 38-footer prowled around Fredonia, Kansas, in July 1860, while residents of Ligonier, Pennsylvania, offered a $3,000 reward for capture of the 35-footer that haunted their town a month later. During the same year, a 50-foot snake frightened livestock in Warren County, Illinois. Farmers at Prairie Grove, Kansas, killed a 15-foot serpent in 1861. Four years later, "anacondas" laid siege to Ohio, including a stovepipe-thick specimen in 1865, a 30-footer in 1868, and two separate snakes - 11 feet 2 inches and 16 feet - killed in 1869. Illinois reported a 33-footer killed "with great difficulty" in May 1866. A giant 28-30 feet long caused "a sensation" in Canaan, Connecticut, in 1867. The busy year 1868 brought reports from California (40-60 feet), New Hampshire (23 feet), and Tennessee (35-45 feet). Ohio's specimens of 1869 were mere striplings, beside the 38-footer seen in Kansas that year.[20]

The 1870s produced so many sightings - 36 in all, from 22 U.S. states and Ontario, Canada - that space restrictions preclude describing each event. Ohio led the field with seven reports during 1871-79; Pennsylvania logged four between 1870 and 1875; California and Missouri claimed three each; Illinois, Maryland and New York each had two incidents; single reports emerged from Connecticut, Florida, Indiana, Kansas, Louisiana, Maine, Mississippi, North Carolina, Ontario, Tennessee, Utah and Vermont. Record sizes for the busy decade belong to a Kansas specimen 37 feet 9 inches long, Indiana's horned monster measuring 34 feet 3 inches, a New York serpent measuring 30-40 feet, and a 31-footer from California. Reptiles 15-20 feet long were almost routine.[21]

The 1880s produced 39 giant-snake reports from 18 U.S. states and Ontario. Ohio led the nation once again, with seven reports; Pennsylvania topped its 1870s record with five; California, Georgia and West Virginia logged three sightings each; Illinois and Ontario each produced two; single incidents occurred in Iowa, Kentucky, Maine, Maryland, Michigan, Mississippi, Missouri and New Jersey. In general, the snakes observed were larger than those from the previous decade, including three of 50 feet or longer, seen in California (1884) and Illinois (1882, 1885). A slightly smaller California specimen, 43 feet 7 inches long, was killed after attacking Chinese laborers in 1882. West Virginia ranchers stoned a 40-footer that had swallowed several sheep near Milton, in 1886.[22] Forty-eight monster serpent reports exist for the 1890s. Ohio leads again, with 11; Indiana boasted 7; Pennsylvania logged 5; Illinois and Missouri claimed 4 each; Iowa produced 3; California, Connecticut, Florida, and New York each filed 2; single sightings came from Kansas, Maine, Maryland, Michigan, Ontario, Virginia and Wisconsin. Ohio's 100-foot reptile from 1890 holds the continental record, while 50-footers were seen in Illinois (1894), California (1895), and Kansas (1897).[23]

The first decade of the 20th century was relatively quiet, producing 29 sightings from 13 states. Pennsylvania took the lead, with 10 reports, while Ohio followed closely with 8. Iowans reported two encounters, while single sightings emerged from Alabama, California, Florida, Georgia, Indiana, Michigan, Nebraska, Tennessee, and Wisconsin. The reptiles seen were smaller, too, with none reported larger than 30 feet.[24]

Sightings declined again between 1910 and 1919, with only 23 reported from 12 states. Pennsylvania dominated the decade with nine reports; Indiana, Maryland, Montana and Ohio each produced two; while single sightings came from California, Florida, Iowa, Kentucky, Minnesota and North Carolina. Thirty-footers were seen in 1910 and 1913, with 20-footers in 1911 and 1912, but the rest were relatively small.[25]

The 1920s produced 18 sightings from 8 states. Indiana led the field with six reports, while Pennsylvania contributed five. Ohio logged two sightings, and individual cases emerged from California, Kentucky, Missouri, New York and South Carolina. Indiana's 35-foot monster of 1920 was the decade's largest reptile.[26]

The 1930s brought a depression to North America - along with 18 giant-snake sightings from 9 states. Pennsylvania residents reported seven monster reptiles; Kansas claimed three; Indiana and Ohio harbored two apiece; while single sightings came from California, Florida, Iowa, Nebraska and North Carolina. The largest reported serpents were 25-footers seen in Kansas (1933, 1935) and California (1934).[27]

The 1940s, encompassing World War II, produced only 12 serpent sightings from 6 states. Indiana claimed half the total, while Ohio logged two reports and single sightings emerged from Georgia, Michigan, Minnesota and Texas. The largest snakes reported were an Indiana 25-footer (1946) and a 22-footer in Minnesota (1941).[28]

The Cold-War 1950s produced a slight upswing in sightings, with 16 reported from 10 states. Pennsylvania recaptured its dominance with five reports; Alabama and Indiana logged two each; with individual sightings filed from Arizona, Colorado, Florida, Michigan, North Carolina, Oklahoma and Texas. Pennsylvania claimed a 40-foot serpent in 1959, while 30-footers were seen in Florida (1950) and Alabama (1957, 1959).[29]

As importation of exotic reptiles became increasingly popular, from 1960 through the end of the 20th century, unexplained giant-snake sightings *declined* rapidly. The 1960s produced only nine cases from eight states, including Arkansas, Florida, Kansas, Kentucky, Maryland, Michigan and Pennsylvania. The largest specimens sighted, both 30-footers, were seen in 1962 - one each in Florida and Kentucky. Six sightings from three states mark the 1970s, including a 30-foot reptile seen in Montana during 1978. The 1980s offer only two cases, one each from Pennsylvania and Texas, while Pennsylvania contributes the only mysterious case on file from the 1990s.[30]

By the dawn of the 21st century, Burmese pythons had established breeding populations in Florida, but unsolved giant-snake sightings lay elsewhere, with five cases reported from Arkansas, Illinois, Pennsylvania and South Dakota during 2000-07. The largest specimen reported, a 17-footer in Pennsylvania, was puny beside the monsters reported in prior decades.[31]

Giant Rattlers

Rattlesnakes are unique to the Western Hemisphere, where herpetologists recognize at least 33 species and 68 subspecies. The largest known species is the eastern diamondback (*Crotalus adamanteus*), with a record length of 8 feet 3 inches. Next largest are the western diamondback (*Crotalus atrox*) at 7 feet 8 inches and the timber rattler (*Crotalus horridus*) at 6 feet 2.5 inches.[32] Nonetheless, reports of much larger rattlers - including some truly huge specimens - have been logged throughout North America between Colonial times and the late 1950s.

As uncovered in archival work by Chad Arment and Jerome Clark, the giant rattlers include:

- 1714: A mention of multiple 17-foot rattlers by author Laurence Klauber, who fails to note their location.[33]
- 1750: A reference to 18-footers, again by Klauber, with no location cited.[34]
- 1753: A 22-foot specimen cited by Klauber, again with no location.[35]
- An undated "early" report of a 21-foot rattler killed in Oklahoma, again published by Klauber. French explorers "discovered" Oklahoma in the 18th century; it became a state in 1907.[36]
- 1857: A 20-foot rattler killed in Harlan County, Kentucky.[37]
- 1858: A 17-foot rattler killed by a "tiger hunter" in Medina County, Texas.[38]
- 1859: A rattler 21 feet 6 inches long, with 103 segments in its rattle, killed near Danville, Illinois.[39]
- 1870: A rattler 12 feet long and 18 inches in diameter, seen in Crawford County, Indiana.[40]
- 1877: An 18-foot specimen killed near Eufala, Oklahoma.[41]
- 1878: A rattler 14 feet 9 inches long, with 1.5-inch fangs, killed in Dooly County, Georgia, in May.[42]
- 1878: An unmeasured rattler weighing 125 pounds, with 42 segments to its rattle, killed in August near Independence, Missouri.[43]
- 1878: A rattler 15 feet 6 inches long and "as large around as a big blue bucket" killed in Brevard County, Florida, in November.[44]
- 1879: A specimen 14 feet 2 inches long, killed in eastern Arizona's White Mountains.[45]
- 1883: A rattler 9 feet 7 inches long, killed near Rocky Hill, Kentucky.[46]
- 1887: A "monster rattlesnake" 12 feet 9¾ inches long, with a body "as large as that of a man of average size," killed in Cherokee County, Oklahoma.[47]
- 1889: The skeleton of a 19-foot rattler with ribs "the size of a small pig's" and 17 segments to its rattle, the largest six inches in diameter, unearthed in Lorain County, Ohio. Incredibly, within the snake's rib cage lay skeletal human remains "of tremendous stature."[48]
- 1890: A rattler exceeding 11 feet, killed in Lee County, Georgia.[49]
- 1891: A 19-foot specimen killed near Columbus, Indiana.[50]
- 1892: Isaac Daines claimed sightings of a 60-foot serpent in Horseshoe Pond, near Vincennes, Indiana. The *Vincennes Commercial Weekly* illustrated its report of the apparent lake monster with a drawing of a huge rattler chasing a man on land.[51]
- 1893: A 17-foot rattler with 164 segments to its rattle and a body as thick as a nail keg, killed

near Salina, Utah.[52]

- 1896: A "huge rattler" that choked to death in Hancock County, Illinois, while trying to swallow "a squirrel and a tartar at the same time."[53]
- 1897: The "Madison County Terror," described by Indiana farmers as a 20-foot rattler, two feet in diameter.[54]
- 1902: Timber prospectors killed a rattler 12 feet 6 inches long in Hardin County, Texas.[55]
- 1904: Road workers killed a 19-foot rattler weighing 127 pounds near Broken Bow, Oklahoma.[56]
- 1908: Indiana farmer John Bascomb killed "Big Jim," a rattler 10 feet 5 inches long which had reportedly terrorized Knox and Sullivan Counties since 1882.[57]
- 1912: Joseph Sponselor shot "the largest rattlesnake ever killed in the United States" on Arizona's White Mountain Indian Reservation. No length was cited, but Sponselor claimed the snake's body was the same thickness as a man's thigh, with 103 segments forming a 12-inch rattle.[58]
- 1917: Hunters in Crittenden County, Arkansas, stalked a 12-foot rattler with a foot-long rattle.[59]
- 1921: Fresh reports of "Big Jim" from Knox County, Indiana, suggest that the monster rattler's 1908 obituary was premature.[60]
- 1923: Surveyors found the skeleton of a 10-foot rattler in a cave near Junction City, Louisiana. Local old-timers claim the snake terrorized their region in the 1860s.[61]
- 1927: Railroad workers killed a rattler 11 feet long, weighing 113 pounds, in Kendall County, Texas.[62]
- 1940s: Laurence Klauber, writing in 1956, related the story of a rattler 25-30 feet long, seen in Baja California "for a number of years" before his book was published.[63]
- 1954: Adrian Gwin, a reporter for the *Daily Mail* in Charleston, West Virginia, described a rattlesnake skin 12 feet 7 inches long and 15 inches wide displayed in Logan County. Allowing for a 30-percent stretch factor during tanning, the live snake would still have been 9 feet 6 inches long.[64]
- 1957: A teenage driver and his passengers reported a near-miss with a rattler "as big around as a log," near Henryetta, Oklahoma.[65]

Proof Positive?

Most giant-snake stories rest solely on anecdotal evidence, but North American archives yield 88 reports of oversized reptiles killed, found dead, or captured alive in the 187 years between 1795 and 1982. The following list does not include recent cases of recognized exotic species killed or captured, such as the 201 pythons caught in Florida during 2002-05.

- May 1793: "Anaconda" killed in Ohio, 26 feet 7 inches.[66]
- July 1843: Dark-green snake killed in Massachusetts, 14 feet 3 inches.[67]
- September 1857: 20-foot rattler killed in Kentucky.[68]
- June 1859: Rattlesnake killed in Illinois, 21 feet 6 inches.[69]
- May 1866: Black snake killed in Illinois, 33 feet.[70]
- July 1868: Unknown spotted snake killed in Tennessee, 29 feet 8 inches.[71]
- August 1868: "Water snake" killed in New Hampshire, 23 feet.[72]
- July 1869: "Immense snake" killed in Ohio, 16 feet.[73]
- October 1869: Black snake killed in Ohio, 11 feet 2 inches.[74]
- February 1871: Black-and-blue snake with yellow spots killed in Kansas, 37 feet 9 inches.[75]

- July 1875: "Bull snake" killed in Missouri, 12 feet 2.5 inches, 180 pounds.[76]
- July 1875: "Horse-whip snake" with "plaited" body killed in Tennessee, 11 feet long.[77]
- August 1875: "Dark" snake with striped head killed in Indiana, 12 feet.[78]
- July 1877: 18-foot rattler killed in Oklahoma.[79]
- November 1877: Unknown 20-foot snake killed in California.[80]
- January 1878: "Boa constrictor" with black and yellow spots killed in Louisiana, 31 feet.[81]
- May 1878: Greenish-coloured 31-foot snake killed in California.[82]
- May 1878: Rattler 14 feet 9 inches long killed in Georgia.[83]
- August 1878: Ohio farmers find a 9-foot newly-shed snakeskin marked with large spots.[84]
- August 1878: 125-pound rattler killed in Missouri.[85]
- November 1878: Rattler 15 feet 6 inches long killed in Florida.[86]
- March 1879: 21-foot snake skeleton with 6-inch ribs found in Connecticut.[87]
- August 1879: Indiana farmer Jacob Rishel kills a snake 34 feet 3 inches long, with 42-inch "tentacles" on its head.[88]
- August 1879: "Monster snake" 21 feet 2.5 inches long killed in Ohio.[89]
- September 1879: Unknown snake killed in Connecticut, 11 feet 5 inches.[90]
- September 1879: Rattler as thick as a man's leg, with a foot-long rattle, killed in Arizona.[91]
- October 1879: 18-foot snake killed in Illinois while trying to swallow a 3-year-old child.[92]
- 1879: Newly-shed snakeskin found in Pennsylvania, 12 feet 6 inches.[93]
- May 1880: "Black or blue racer" killed in Pennsylvania, 14 feet and ½ inch.[94]
- March 1881: Unknown snake killed in Ontario, 16 feet 2 inches.[95]
- June 1881: "Water snake" killed in Georgia, 8 feet 2 inches.[96]
- January 1882: Unidentified snake killed after attacking California fisherman, 43 feet 7 inches.[97]
- June 1883: Rattler 9 feet 7 inches long killed in Kentucky.[98]
- July 1883: Black snake 14 feet 3 inches long killed in Pennsylvania.[99]
- September 1883: Black snake killed in Missouri, 13 feet 11 inches.[100]
- August 1885: Indiana hunters kill a snake "nearly 18 feet" long.[101]
- June 1887: Elias Moser kills a bluish-black "king snake" in Pennsylvania, 16 feet 6 inches.[102]
- July 1887: Ohio hunters kill a snake 17 feet 2 inches long.[103]
- December 1887: Oklahoma railroad workers kill a rattler 12 feet 9¾ inches long.[104]
- July 1888: Black snake killed in Pennsylvania, "14 feet and inches [sic] in length."[105]
- September 1888: 16-foot "pine snake" killed by an Ohio policeman.[106]
- April 1889: Ohio workmen excavate the skeleton of a 19-foot rattlesnake, containing the bones of a "giant man."[107]
- August 1889: Snake "of the moccasin variety" killed in Mississippi, 16 feet 5 inches.[108]
- December 1890: Snake 19 feet 8 inches long caught alive at Lake Kenosia, Connecticut.[109]
- December 1890: Rattler "a little over 11 feet long" killed in Georgia.[110]
- May 1891: 19-foot rattler killed in Indiana.[111]
- October 1891: Unidentified green snake with red spots killed in Illinois, 9 feet 2 inches.[112]
- July 1892: Two 8-foot black snakes killed in Ohio.[113]
- August 1893: Black snake 11 feet 9 inches long killed in New York, while attacking lambs.[114]
- August 1894: Pennsylvania "bark peelers" kill two black snakes: one measures 14 feet 9.5 inches, the other 14 feet.[115]

- September 1894: Unidentified 14-foot snake killed in Ohio.[116]
- July 1895: 31-foot "anaconda" killed in Ohio.[117]
- July 1896: 9-foot rattler as thick as a man's thigh found dead in Illinois.[118]
- April 1897: Python 34 feet 6 inches long caught alive in Florida, by a Smithsonian Institution employee.[119]
- August 1897: Missouri farmer kills a 16-foot snake "very much like a python."[120]
- October 1898: Pennsylvania hunters kill "a number" of unidentified snakes, 8-10 feet long.[121]
- August 1899: Ohio farmers kill a "timber black snake," 12 feet 10.5 inches.[122]
- August 1901: "Monster snake" 15 feet 3 inches long killed in Pennsylvania.[123]
- August 1901: Second Pennsylvania snake killed, "over 20 feet" long.[124]
- January 1902: Florida hunter wounds a large snake, later finding its remains partly devoured by scavengers: 20-30 feet long, head 10 inches wide.[125]
- September 1902: "Blacksnake" 11 feet 1 inch long killed in Pennsylvania.[126]
- July 1903: Ohio fisherman kills 9-foot "boa constrictor."[127]
- August 1903: 13-foot "boa constrictor" killed in Ohio.[128]
- September 1904: 19-foot, 127-pound rattler killed in Oklahoma.[129]
- May 1905: 14-foot "coachwhip" snake killed while attacking a Georgia farmer.[130]
- August 1906: 17-foot "blacksnake" caught alive in Pennsylvania.[131]
- November 1906: "Blacksnake" 12 feet long, 33.5 inches in circumference, killed in Pennsylvania.[132]
- August 1908: 16-foot "chicken snake" killed in Indiana.[133]
- May 1909: "Blacksnake" 11 feet 2 inches long killed in Pennsylvania.[134]
- June 1909: Another Pennsylvania "blacksnake" killed, 14 feet 7 inches.[135]
- July 1910: Ohio farmer kills a black snake 9 feet 8 inches long.[136]
- September 1910: 18-foot snake with dark mahogany spots caught alive in Montana.[137]
- May 1911: Pennsylvania railroad worker kills a "blacksnake" 9 feet 4 inches long.[138]
- February 1912: Rattler with a foot-long rattle containing 103 segments killed in Arizona.[139]
- May 1912: 15-foot boa constrictor caught alive in Minnesota.[140]
- July 1912: Black snake 10 feet 10 inches long killed in Pennsylvania.[141]
- July 1915: Black snake 9 feet 2 inches long killed in Pennsylvania.[142]
- September 1918: 9-foot "cow snake" killed in Kentucky.[143]
- August 1919: Black snake "within a fraction of 12 feet" long, thick as a man's arm, killed after attacking a Pennsylvania hiker.[144]
- August 1923: Skeleton of a 10-foot rattlesnake found in Louisiana.[145]
- July 1931: Ohio residents pull an 18-foot "boa constrictor" from Lake Erie.[146]
- April 1935: 11-foot "cobra" weighing 65 pounds caught alive in Pennsylvania.[147]
- May 1944: Dead snake 13 feet 6 inches long, 6 inches diameter, found and displayed in Ohio.[148]
- July 1946: 14-foot "blacksnake" killed in Ohio.[149]
- May 1950: "Boa constrictor" 10 feet long and "the size of a man's leg" caught alive in Florida.[150]
- August 1950: 28-foot snake weighing 250 pounds caught alive in Oklahoma.[151]
- April 1982: Unidentified 25-foot snake shot and measured by a Texas farmer, then disappears overnight.[152]
- Undated: "Early" report of a 21-foot rattler killed in Oklahoma.[153]

Sadly, despite reports that several of the giants killed or captured were dispatched to museums or zoos, no trace of them remains today. Neither were any photographed, alive or dead. Only the journalistic record stands, subject to claims of hoaxes or exaggeration that cannot be answered with hard evidence.

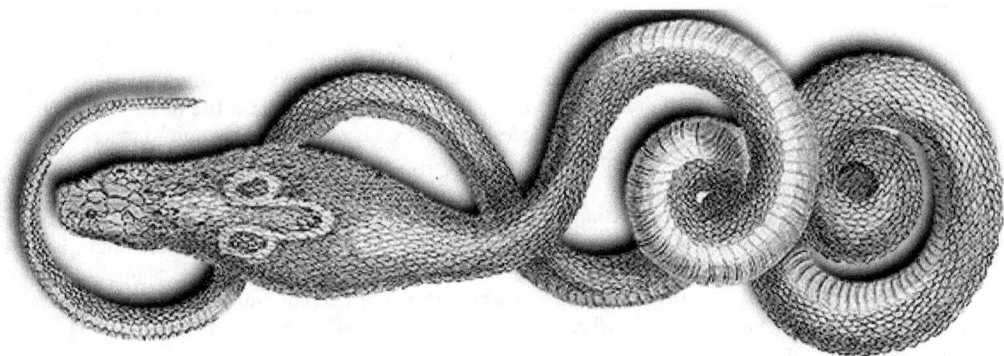

Conclusion

Our journey is concluded, and it ends as it began, in mystery. Having examined tales of giant snakes from every corner of the globe, we still confront the question that began our quest: *Do they exist?*

Clearly, the "Big Six" serpents recognized by science are alive and well in their native habitats on four continents. Trade in exotic reptiles spreads that range worldwide, so sightings of boas and pythons at large in Europe and North America no longer seem surprising.

Still, we are left with hundreds of reports describing snakes much larger than the maximum allowed by modern herpetology, exceeding any snake known from Earth's fossil record. Tales from North America predate the commercial traffic in tropical exotics, and some of the reptiles described resemble nothing known to science.

Five common explanations are advanced for claims of giant snakes: hoaxes; misidentification of known species or other common objects; exotic species at large; "freak" members of known species; and new species unknown to science. We shall examine each in turn, beginning with the most troublesome first.

Hoaxes

Chad Arment identifies three types of giant-snake hoaxes: deliberate frauds or pranks; "tall tales" passed into folklore; and commercial showman's hype. The first includes both individual practical jokes and journalistic fabrications penned to sell newspapers.

While it seems incomprehensible today, some 19th-century newspapers assigned writers to fabricate "monster" reports as a form of entertainment, or to increase sales during circulation slumps. Such reports were especially common in summer, described by editors as "the silly season." Some such hoaxes were belatedly confessed, while others - like the supposed capture of an immature Sasquatch called "Jacko" in British Columbia - endured for decades, misleading generations of serious researchers.[1]

Private hoaxes and pranks are more diverse than media lies. Arment cites the cases of an Alabama youth who created "giant snake tracks" by dragging a watermelon along dusty roads, an Indiana teenager jailed for disturbing the peace after he claimed a near-miss with a snake the size of a fire hose, and the Kansas pranksters who built a monster serpent out of canvas and excelsior in 1931.[2] Logic dictates that other serpent sightings fall into the same deceptive category.

The classic "tall tales" of giant snakes combine mythology with old-fashioned hyperbole, as in frontiersman Davy Crockett's oft-repeated claim that he "kilt a bear" at age three. Such stories might be circulated as pure entertainment, often growing more outlandish as they made the rounds, or might be used as a form of social control - as for frightening delinquent children or keeping trespassers away from private property. Exaggeration via rumors, as where a six-foot snake grows exponentially in retelling the tale, may also play a role in such cases.[3]

Finally, showman's hype lends a hand when carnival barkers promote reptilian sideshow attractions, and may find its way into mainstream journalism in a bid to promote tourism. Some self-styled skeptics claim that Nessie exists only in the imagination of Scottish tourist boards, and the same may be said of certain newspaper reports - as when 19th-century residents of Polk County, Missouri, claimed possession of a snake "842 feet long, and of corresponding girth." More recently, the hype surrounding "Fragrant Flower" added 27 feet to a normal 22-foot python.[4]

Misidentification

Witnesses who are unfamiliar with reptiles often mistake one species of snake for another. Fear may also prompt exaggeration of a serpent's size - or trick a startled mind into mistaking some common, inanimate object for a man-eating monster. Cases in point include the following:

- August 1870: An Indiana "blacksnake" 10-12 feet long proved, on closer examination, to be one snake swallowing another.[5]
- August 1893: Ohio pursuers of a "monster snake" found their quarry to be a burrowing ground-hog.[6]
- January 1921: Another Ohio snake-hunting party pumped bullets into a 40-foot hose, used to siphon water from a pond to a nearby sawmill.[7]
- June 1935: The trail of a "giant snake" in North Dakota was produced by a farmer's bull, dragging a pole rigged on a collar to prevent it from jumping fences.[8]
- July 1939: An escaped dog, dragging its chain, produced more "snake tracks" in Wisconsin.[9]
- October 1966: Louisiana sheriff's deputies and firefighters torched a field, in pursuit of a 15-foot snake. When finally cornered and killed, the reptile measured only 4 feet 11 inches.[10]

Known Exotics

As previously noted, Asian pythons have established breeding populations in Florida. In February 2008 spokesmen for the U.S. Geological Survey projected that the snakes might soon spread nationwide, colonizing one-third of the United States from Virginia to California, infesting all or part of 15 states. Already, reports of large boas and pythons caught above the theoretical USGC cut-off line have been logged from states including Illinois, Indiana, Kansas, Ohio and New York.[11]

Today, we accept sale of exotic reptiles as routine, along with their occasional escape or release by negligent owners. But, in fact, newspaper records document at least 25 cases of supposed exotic giant snakes at large in the U.S. between 1836 and 1954. Those cases include:

- October 1836: The Canadian steamship *Royal Tar* sank while carrying a circus troupe from New Brunswick to Portland, Maine. Maine residents blame escaped reptiles for later sightings of large snakes on offshore islands.[12]
- August 1865: Ohio residents searched in vain for an anaconda as thick as a stove pipe.[13]
- July 1875: A 20-foot anaconda was caught near Baltimore, Maryland, then escaped and vanished.[14]
- June 1880: Reports of an 18-foot snake in Kentucky cite "a tradition that such a reptile escaped from a circus ...many years since and still lives there."[15]
- December 1891: Virginia authorities announce that a dead 15-foot snake was "evidently an anaconda which had probably escaped" from an unnamed traveling show.[16]
- September 1894: Reporters claim that two "snakes as big as logs," seen in Iowa, are boa constrictors escaped from the traveling Van Amberg Circus.[17]
- July 1895: Ohio residents kill a 31-foot anaconda which allegedly escaped from a circus in 1887.[18]
- September 1895: Iowans search for a 14-foot snake "as big around as a pail," claiming it and several others were released at Davenport by its owner, the late "Dr. Wood."[19]
- April 1897: Walter Ralston of the Smithsonian Institution identified a 34½-foot snake caught in Florida as a python of unknown origin.[20]
- October 1901: An African rock python is caught in Pennsylvania. Reports claiming "the liberation of large snakes" remain unsubstantiated.[21]
- August 1903: An Ohio firefighter kills a 13-foot boa constrictor which "is supposed to have escaped from a traveling show."[22]
- June 1911: Indiana farmers suspect that a 15-foot snake which has raided their hog pens for three consecutive years is a "derelict from a circus."[23]
- May 1912: A 15-foot boa "thought to have escaped from a circus last summer" is caught in Minnesota.[24]
- February 1921: Members of a traveling carnival recapture a boa constrictor that escaped in Kentucky some time earlier.[25]
- June 1926: Residents of Hayward, California, report sightings of a 10-foot boa which "escaped from a circus as a baby," years earlier.[26]
- September 1931: Pennsylvania forest rangers claim that an "enormous snake" seen in Mont Alto State Park "may be" a python that escaped from an unknown private owner.[27]
- May 1933: Pennsylvania's New Castle *News* opines that another giant snake "might be a boa constrictor or a python escaped from a circus."[28]
- September 1941: The *Daily Dispatch,* of Brainerd, Minnesota, reports a 22-foot python at large, noting unsubstantiated "rumors" of its escape from a carnival.[29]
- June-July 1944: Police and vigilantes search in vain for Ohio's 18-foot "Peninsula Python."[30]
- August 1946: Indiana residents mount fruitless searches for a supposed boa constrictor or python.[31]
- March 1950: Ranchers claim that a 19-foot boa constrictor "which has survived three of Colourado's sub-zero winters" is preying once more on their cattle.[32]
- May 1950: Snake-hunters from Jacksonville, Florida, capture a 10-foot boa constrictor described as the "grandchild" of a 30-foot reptile "believed to have escaped from a circus train 53 years ago."[33]

- August 1950: Oklahoma hunters seek a boa constrictor "as long as two trucks and thick as a man," seen around Holdenville since 1935.[34]
- October 1954: Texas residents search in vain for "Python Pete," an 18-footer that escaped from Fort Worth's zoo.[35]

From these accounts, it would appear that Third World reptiles may have colonized North America more than a century before their first beachhead was formally acknowledged - and their range has not been limited to southern states with year-round warm weather. Meanwhile, the majority of giant-snake reports - at least, from North America - seem to describe reptiles unknown to science.

Which leaves us with ....

Something Else?

When hoaxes, honest mistakes, and stray exotics are eliminated, giant snakes are generally explained either as "freak" individuals of a known Big Six species, or as members of a species yet unclassified.

The first explanation clearly works best in regions where large boas or pythons normally thrive, though accounts of improbably large rattlesnakes also crop up across North America. We have noted anecdotal evidence of anacondas, pythons and rattlers far longer than the record specimens acknowledged by modern science, yet none have been preserved, either in the flesh or through indisputable photos. Until such time as specimens are killed or captured, skeptics will dismiss reports of 50- or 60-foot pythons and 15- or 20-foot rattlers as pure fantasy.

But if such giants from known species *were* substantiated, how would that affect reports of *other* huge snakes bearing no resemblance to known species? How would a 50-foot anaconda explain sightings of the *sucuriju gigante* two or three times as large? How would proof of oversized rattlers explain Tennessee's spotted 29-footer from 1868? The blue-and-black monster with yellow spots caught in Kansas, in 1871? The 31-foot specimen with black and yellow spots, killed in Louisiana in 1878? The green snake with red spots, killed in Illinois during 1891? Or the series of immense "blacksnakes" killed in Pennsylvania since the 1870s?[36]

Researcher Chad Arment admits the possibility of "small localized populations of one or more unknown species" in North America, while deeming it unlikely. Factors weighing against that explanation for giant-snake sightings in the U.S. and Canada include a lack of fossil evidence for unknown snake species and the impact of cold winters on reptiles above the Mason-Dixon Line.[37]

Author Darren Naish contends that since "no-one has yet produced a snake that exceeds 10 m [32 feet 6 inches] in total length" - a reticulated python reportedly shot in 1912, but not preserved - reports of larger snakes must arise from "rumours and reports of fossil snake bones that have been estimated by some workers to have belonged to truly gargantuan reptiles." Even then, however, Naish allows a maximum length of 9 metres [29 feet 3 inches] for the prehistoric *Gigantophis garstini* of North Africa, which he ranks as "the biggest" of prehistoric contenders. Nonetheless, Argentinean paleontologists Zulma Gasparini, Leonardo Salgado and Rodolfo Coria published a paper in 1993 describing another prehistoric reptile group, the sebecosuchians, described as "huge boids up to 29 meters [94 feet 3 inches!] in length." Further research, however, reveals that the sebecosuchians were, in fact, crocodilians. That confusion notwithstanding, Naish concludes that "snakes exceeding 15 m [48 feet 9 inches] in length almost certainly never existed."[38]

Rational skeptics and knee-jerk debunkers alike naturally refuse to accept the existence of truly giant snakes until such time as irrefutable evidence of their existence is produced. By the same token, the argument sometimes advanced against these and other cryptids - to wit, "They don't exist because they *can't* exist" - is a prime example of fallacious circular reasoning. Large unknown reptiles *may* exist, or may have existed within historical times, despite the fact that none reside today in a museum or zoo.

In short, the mystery endures.

The Boas

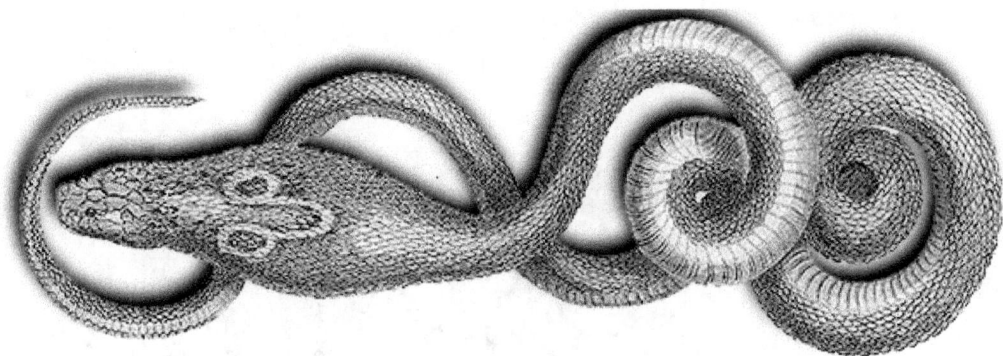

Acknowledgements

I owe special thanks to three persons for their support during production of this manuscript: To Jon Downes at CFZ Press for his unflagging enthusiasm for the project; to Dave Frasier at Indiana University, for retrieving several rare articles on the subject; and to my wife Heather for her generous review of the work. Finally, my own work would have been impossible without the Herculean archival research performed by Chad Arment, Jerome Clark, Mark Hall, and the late Bernard Heuvelmans. Every effort has been made to locate copyright holders for the illustrations used in this work. Most fall in the public domain; diligent searches for owners of the rest - which appear without credit lines in various other publications and across the World Wide Web - proved fruitless at press time for the work in hand.

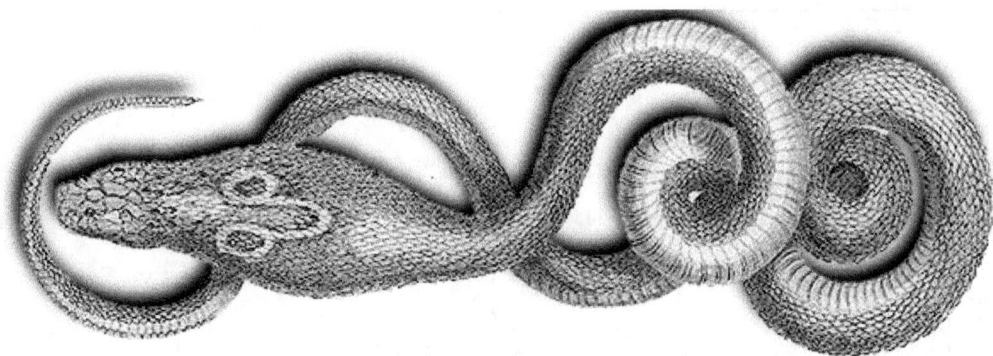

Notes

Chapter 1
1. Wikipedia, "Serpent (symbolism)."
2. Internet Movie Database.
3. Internet Movie Database.
4. Mattison, p. 18.
5. Mattison, p. 18; Wikipedia, "Anaconda."
6. Mattison, pp. 19-20; Wikipedia, "Python reticulatus."
7. Mattison, p. 20; Wikipedia, "Python sebae."
8. Mattison, p. 20; Wikipedia, "Python molurus."
9. Mattison, p. 22; Wikipedia, "Morelia amethistina."
10. Mattison, p. 22; Wikipedia, "Boa constrictor."
11. Twigger, pp. 12-13; David Gordon, "The search for the $50,000 snake." Encarta.
12. Arment, pp. 44-5; Wikipedia, "List of European reptiles."
13. Arment, pp. 10-43.
14. Arment, pp. 44-64.
15. Arment, p. 66; Newton, *Florida's Unexpected Wildlife,* pp. 4-6.

Chapter 2
1. Wikipedia, "List of European reptiles."
2. "Why Ireland has no snakes," Smithsonian National Zoological Park, http://nationalzoo.si.edu/ Animals/ ReptilesAmphibians/NewsEvents/irelandsnakes.cfm.
3. Costello, pp. 183-4.
4. Ibid., p. 184.
5. Wikipedia, "List of European reptiles."
6. Coghlan, *Dictionary,* p. 29; Eberhart, p. 44; Shuker, *Beasts,* pp. 185-6.
7. Freeman, *Explore Dragons,* p. 32.
8. Cohglan, *Dictionary,* p. 214; Freeman, p. 30.
9. Freeman, *Explore Dragons,* p. 5.
10. Ibid., p. 26.
11. Ibid., p. 11.
12. Ibid., pp. 16-17.
13. Ibid., pp. 23-4.

14. Ibid., p. 26.
15. Ibid., p. 28.
16. Ibid., pp. 12-13; Shuker, *Dragons,* pp. 12-15.
17. Freeman, *Explore Dragons,* p. 28.
18. Ibid., p. 13.
19. Coghlan, *Cryptosup,* p. 4
20. Freeman, *Explore Dragons,* pp. 33-34; Eberhart, pp. 37-8; Karl Shuker, "A water vampire." *Fate* 43 (March 1990): 86-88.
21. Croghlan, *Dictionary,* p. 29; Cryptozoologix, http://www.angelfire.com/pq/cryptozoologix/beithir. html.
22. Beastwatch UK.
23. Ian Whadcoat, "Snakes alive!" *Leicester Mercury*, 29. Feb. 2000.
24. *Northern Echo,* 17 Nov. 2000.
25. Beastwatch UK.
26. Ibid.
27. BBC News, 3 September 2002.
28. Beastwatch UK; BBC News, 2 Oct. 2002.
29. *Birmingham Evening Mail,* 22 Jan. 2003.
30. Beastwatch UK.
31. *Swindon Evening Advertiser*, 18 Feb. 2004.
32. *Daily Record,* 16 March and 16 April, 2004.
33. *The Scotsman,* 11 Aug. 2004.
34. Beastwatch UK.
35. BBC News, 1 March 2005.
36. BBC News, 2 Aug. 2005.
37. BBC News, 13 Aug. 2005.
38. BBC News, 6 Sept. 2005.
39. Ananova News, 18 Oct. 2005.
40. BBC News, 9 April 2007.
41. BBC News, 17 April 2007.
42. BBC News, 13 June 2007.
43. BBC News, 25 June 2007.
44. Wikipedia, "List of European reptiles."
45. Rose, pp. 246-7.
46. Clark, *Unexplained!,* p. 440; Freeman, pp. 62-3.
47. Clark, *Unexplained!,* pp. 440-1; Freeman, p. 63.
48. Clark, *Unexplained!,* pp. 441-2.
49. Sven Rosen, "The dragons of Sweden," *Fate* 35 (April 1982): 36-45.
50. Bord and Bord, pp. 323-4.
51. Bord and Bord, p. 249; Eberhart, p. 324.
52. Bord and Bord, p. 352; Eberhart, p. 324.
53. Coghlan, *Dictionary,* p. 21.
54. Boingboing, http://www.boingboing.net/2007/10/16/ french-lady-finds-py.html.
55. Gould, p. 165.
56. Gould, pp. 187-8; Coghlan, *Cryptosup,* p. 47.
57. Wikipedia, "Dragon"; Bord and Bord, p. 339; Eberhart, p. 112.
58. Bord and Bord, p. 339.
59. Freeman, *Explore Dragons,* p. 68; Gould, p. 165.
60. Gould, p. 166-7.
61. Mystery Animals of Europe, http://www.europacz.com/ EnReptiles.htm.

62. Mystery Animals of Europe.
63. Marcus Scibanicus, "Strange creatures from Slavic folklore." *North American BioFortean Review* 3 (Oct. 2001): 54.
64. Wikipedia, "Dragon."
65. Eberhart, p. 148.
66. Coghlan, *Dictionary,* p. 94.
67. Wikipedia, "List of European reptiles."
68. Gould, p. 170; Wikipedia, "Metrodorus"; Wikipedia, "Pontus."
69. Mystery Animals of Europe.

Chapter 3

1. Peter Alden, Richard Estes, Duane Schlitter and Bunny McBride. *National Audubon Society Field Guide to African Wildlife* (New York: Alfred A. Knopf, 1995), p. 906.
2. Mattison, p. 20; Pope, pp. 15-16; Wikipedia, "Python sebae"; "Facts vs. Myths," AHP Exotics, http://www. ahpexotics.net/WorldRecords.html.
3. Pope, pp. 227-8; "Boys claim friend eaten by snake," *Sydney Morning Herald* (Nov. 22, 2002); "Giant python swallows child in South Africa," *Jet* (Dec. 16, 2002).
4. Wikipedia, "Python regius"; Wikipedia, "Python anchietae."
5. Gould, p. 171.
6. P.G.P. Meyboom, *The Nile Mosaic of Palestrina* (Aylesbury, England: Brill, 1995, p. 225.
7. Wikipedia, "Marcus Atilius Regulus."
8. Freeman, *Explore Dragons,* p. 72; Gould, pp. 173-4; Shuker, *Dragons,* pp. 26-9.
9. Gould, pp. 168, 170, 175.
10. Ibid., 171-2.
11. Freeman, *Explore Dragons,* p. 72.
12. Eberhart, p. 468; Freeman, p. 72.
13. Bord and Bord, p. 306; Eberhart, p. 468; Freeman, pp. 72-3.
14. Eberhart, p. 468; Freeman, p. 73.
15. Heuvelmans, *On the Track,* p. 535.
16. Ibid., pp. 535-6.
17. Ibid., pp. 536-7.
18. Ibid.
19. Ibid., pp. 537-8.
20. Ibid., 538; Mackal, pp. 260-2.
21. Shuker, *Extraordinary,* p. 31.
22. Ibid., p. 32; U.S. Dept. of the Navy, *Poisonous Snakes of the World* (Washington: U.S. Government Printing Office, 1966), pp. 80-1, 91-2, 94-8, 125-6.
23. Eberhart, pp. 115, 388; Shuker, *Extraordinary,* pp. 31-2.
24. Shuker, *Extraordinary,* p. 32.
25. Ibid.
26. Ibid., p. 33; Eberhart, p. 357.
27. Shuker, *Extraordinary,* p. 34; Eberhart, p. 116.
28. Shuker, *Extraordinary,* p. 34.
29. Ibid., pp. 35-6.
30. Ibid., pp. 36-7, 40.
31. Eberhart, p. 386.
32. Mackal, pp. 255-6.
33. Gibbons, pp. 73-4.
34. Mackal, pp. 259-60, 263.
35. Eberhart, p. 444.

36. Gould, p. 178; "Whydah," Wikipedia, http://en. wikipedia.org/wiki/Whydah.
37. Here We Are in Africa, http://www.pythons.net/ largesnake3.html (accessed 2004).
38. Eberhart, p. 444; Gibbons, pp. 91-2.
39. Shadowlands, http://theshadowlands.net/creature. htm#congo.
40. Unexplained Mysteries, http://www.unexplained-mysteries.com/forum/index. php? showtopic=66418.
41. Gibbons, pp. 92-3; "French Equatorial Africa," Wikipedia, http://en.wikipedia.org/wiki/ French_Equatorial_ Africa; "Sangha River," Wikipedia, http://en.wikipedia.org/ wiki/Sangha_River.
42. Coghlan, *Further*, p. 32; "South African Cryptids," About.com, http://paranormal.about.com/library/ blstory_ august05_30.htm.
43. "South African 'horse-headed snake,'" *Cryptozoology Review* 4 (Summer 2000): 4; Eberhart, p. 511.
44. The Wild Coast, http://www.wildcoast.org.za/wc.

Chapter 4

1. "Asia," Wikipedia, http://en.wikipedia.org/wiki/Asia.
2. "Python reticulatus," Wikipedia, http://en.wikipedia. org/wiki/Reticulated_python; Pope, pp. 16, 161-3.
3. "World's largest snake caught," Ananova (29 Dec. 2003), http://www.ananova.com/news/story/sm_850885.html?menu=; Tim Moore, "Zoo's monster python could be off the scale for size," *The Times* (London), 30 Dec. 2003; John Aglionby, "Captured python said to be world's biggest snake," *The Guardian* (London), 30 Dec. 2003; "Stay still, will you?" *The Guardian,* 4 Jan. 2004; "Reports of monster snake exaggerated," *Arizona Republic,* 13 March 2004.
4. Pope, p. 227; "Python reticulatus," Wikipedia.
5. "Python reticulatus," Wikipedia; Silvio Bruno, "I serpenti giganti," *Criptozoologia* 4 (1998): 16-29; Ian Steward, "Giant python killed after trying to swallow man," *The Star* (Malaysia), 16 Sept. 1995; "Woman dies, swallowed by python," *Oakland* (CA) *Tribune,* 22 Nov. 2003.
6. "Man Eating Snakes," http://arachnophiliac.info/burrow/ oddities_man_eating_snakes.htm; "Snake Food," Snopes.com, http://www.snopes.com/horrors/animals/anaconda.asp; "Man Eaten by Python," http://clipmarks.com/clipmark/F7EDE23C-DB56-4ED1-AB07-2756812560F7.
7. Pope, p. 161; "Python molurus," Wikipedia, http://en. wikipedia.org/wiki/Python_molurus; "Burmese python," National Geographic, http://animals.nationalgeographic.com/ animals/reptiles/burmese-python. html; "Burmese python," Wikipedia, http://en.wikipedia.org/wiki/Python_molurus_ bivittatus.
8. "King cobra," National Geographic, http://animals. nationalgeographic.com/animals/reptiles/king-cobra.html; "King cobra," Wikipedia, http://en.wikipedia.org/wiki/ King_Cobra.
9. "Siberia," Wikipedia, http://en.wikipedia.org/wiki/ Siberia.
10. "Siberia," Wikipedia.
11. "Siberia," Wikipedia; "Primorsky Krai," Wikipedia, http://en.wikipedia.org/wiki/Primorsky_Krai.
12. Paul Stonehill Bio, http://www.coasttocoastam.com/ guests/986.html; Paul Stonehill, "Giant serpents of the Russian Far East," *Strange Magazine* no. 13 (Spring 1994): 29; Paul Stonehill, "The Russian Snowman," http://www. bigfootencounters.com/creatures/russian.htm.
13. "Mongolia," Wikipedia, http://en.wikipedia.org/wiki/ Mongolia; "Gobi Desert," Wikipedia, http://en. wikipedia.org /wiki/Gobi_Desert.
14. Freeman, *Explore Dragons*, p. 2.
15. Ibid.
16. Ibid, pp. 73-5.
17. Gould, pp. 178-9, 234-5, 241; "Guangxi," Wikipedia, http://en.wikipedia.org/wiki/Guangxi; "Athanasius Kircher," Wikipedia, http://en.wikipedia.org/wiki/Athanasius_Kircher; "Python," Wikipedia, http://en.wikipedia.org/wiki/Python_ %28genus%29.
18. Gibbons, p. 121.

19. Ibid.

20. Ibid., p. 122.

21. "Python molurus," Wikipedia; Gibbons, pp. 122-3.

22. Gould, p. 169.

23. Ibid., pp. 169-70.

24. Ibid., p. 170; "Megasthenes," Wikipedia, http://en. wikipedia.org/wiki/Megasthenes.

25. "Python molurus," Wikipedia; "Spitting cobra," Wikipedia, http://en.wikipedia.org/wiki/ Spitting_cobra.

26. Burma, Wikipedia, http://en.wikipedia.org/wiki/Myanmar; "Python," Wikipedia; "King cobra," Wikipedia.

27. Alan Rabinowitz, *Beyond the Last Village* (Washington: Island Press, 2001), p. 116; "Putao," Wikipedia, http://en. wikipedia.org/wiki/Putao.

28. "Sherard Osborn," Wikipedia, http://en.wikipedia.org/ wiki/Sherard_Osborn; "Kedah," Wikipedia, http://en. wikipedia.org/wiki/Kedah; Gould, pp. 175-6.

29. "Boa constrictor," Wikipedia, http://en.wikipedia.org/ wiki/Boa_constrictor; "Python," Wikipedia; "King cobra," Wikipedia.

30. Gould, pp. 176-8.

31. "Bera Lake," Wikipedia, http://en.wikipedia.org/wiki/ Bera_Lake; Tasik Bera, http://www.marimari. com/content/ malaysia/popular_places/lakes/bera/bera.html; "Pahang," Wikipedia, http://en.wikipedia. org/wiki/Pahang.

32. Dinsdale, pp. 101-4.

33. Ibid., p. 104; Shuker, *In Search,* p. 34.

34. Freeman, *Explore Dragons,* p. 84; "Semenyih," Wikipedia, http://en.wikipedia.org/wiki/Semenyih; "Selangor," Wikipedia, http://en.wikipedia.org/wiki/Selangor.

35. "Singapore," Wikipedia, http://en.wikipedia.org/ wiki/Singapore.

36. "John Frederick Adolphus McNair," Wikipedia, http://en. wikipedia.org/wiki/ John_Frederick_Adolphus_McNair.

37. Gould, p. 176.

38. Coghlan, *Dictionary,* p. 174; Eberhart, p. 365; Freeman, *Explore Dragons,* p. 81; "Mekong," Wikipedia, http://en. wikipedia.org/wiki/Mekong.

39. Eberhart, p. 365; Freeman, *Explore Dragons,* pp. 5-7; Freeman, "In the coils."

40. "The fantastic creatures of Angkor"; Freeman, "In the coils"; Pennapa Hongthong, "Naga fireballs: Laos to cash in on show," *The Nation* (Bangkok), 9 Oct. 2003; Napanisa Kaewmorakot, "Science Ministry solves Naga fireballs mystery," *The Nation,* 12 Oct. 2003; "Ministry stands by fireball hypothesis," *The Nation,* 15 Oct. 2003.

41. Freeman, *Explore Dragons,* pp. 83-4.

42. Ibid., pp. 82-3; Freeman, "In the coils."

43. Freeman, *Explore Dragons,* pp. 81-2.

44. Freeman, "In the coils."

45. Freeman, *Explore Dragons,* p. 82; Freeman, "In the coils."

46. Freeman, *Explore Dragons,* p. 83; Freeman, "In the coils"; "Madtsoiidae," Wikipedia, http://en. wikipedia.org/ wiki/Madtsoiidae; "Gigantophis garstini," Wikipedia, http://en.wikipedia.org/wiki/ Gigantophis_garstini.

47. "Giant Snake photo, Vietnam 1964," Unexplained Mysteries Discussion Forum, http://www. unexplained-mysteries.com/forum/lofiversion/index.php/t61441.html.

48. Freeman, *Explore Dragons,* p. 83; "Kong Le & the dragon," *Time,* 21 Oct. 1966; "Kong Le," Wikipedia, http://en.wikipedia.org/wiki/Kong_Le.

49. Gould, p. 178; "Dutch East Indies," Wikipedia, http:// en.wikipedia.org/wiki/Dutch_East_Indies.

50. Gould, pp. 176, 178; "William Broderip," Wikipedia, http://en.wikipedia.org/wiki/ William_Broderip; "Sulawesi," Wikipedia, http://en.wikipedia.org/wiki/Sulawesi.

51. Gould, p. 178.

52. Heuvelmans, *In the Wake,* pp. 272, 382.

53. Ibid., pp. 382-3.

54. Ibid., pp. 382, 579.

55. Newton, *Encyclopedia of Cryptozoology,* p. 5.

56. "Borneo," Wikipedia, http://en.wikipedia.org/ wiki/Borneo; Elaine Engeler, "52 new species discovered on Borneo island," Associated Press, 18 Dec. 2006; "New cat species discovered in Borneo," Mongabay.com (14 March 2007), http://news.mongabay.com/2007/0314-leopard.html.

57. Richard Freeman, "The Orang-Pendek," *Fortean Times* 208 (April 2006), http://www.forteantimes.com/features/ articles/103/the_orangpendek.html; "Sumatra," Wikipedia, http://en.wikipedia.org/wiki/Sumatra.

58. Gould, pp. 167-8; "Abu al-Hasan 'Alī al-Mas'ūdī," Wikipedia, http://en.wikipedia.org/wiki/Masudi; "Al Idrisi," http://www.geocities.com/pieterderideaux/ idris.html; "Taprobana," Wikipedia, http://en.wikipedia.org /wiki/Taprobana.

59. "Philippines," Wikipedia, http://en.wikipedia.org/ wiki/Philippines.

60. Gould, p. 175; Japanese Holdouts: Registry, http://www.wanpela.com/holdouts/registry.html#phil; Wikipedia, "Mindoro."

Chapter 5

1. Wikipedia, "Oceania"; Wikipedia, "Melanesia"; Wikipedia, "Bismarck Archipelago"; Wikipedia, "Micronesia"; Wikipedia, "Polynesia."

2. Immigration New Zealand, http://www.immigration.govt.nz/ nzopportunities/aboutnz/faq/spiders.htm; Wikipedia, "Reticulated python"; Wikipedia, "Python molurus."

3. Wikipedia, "Australia"; Wikipedia, "List of islands by area"; Dangerous Snakes of Australia, http://home.iprimus. com.au/gunnado/snakes.html.

4. Pope, pp. 163-4; Wikipedia, "Morelia amethistina."

5. Wikipedia, "Morelia (genus)."

6. Wikipedia, "Antaresia"; Wikipedia, "Aspidites."

7. Eberhart, p. 455; Rose, pp. 305-6.

8. Wikipedia, "Taipan"; Eberhart, pp. 455-6.

9. Gould, pp. 179-80; Eberhart, p. 336.

10. Gould, p. 180.

11. Ibid.

12. Ibid.

13. Wikipedia, "New Guinea"; Papuatrekking.com, http://www. papuatrekking.com; Wikipedia, "Ropen."

14. Wikipedia, "Morelia (genus)"; Wikipedia, "Apodora"; Wikipedia, "Leiopython."

15. Chris Bishop, *The Encyclopedia of Weapons of World War II* (New York: Barnes & Noble, 1998), pp. 99-100.

16. Wikipedia, "Morelia (genus)."

17. Wikipedia, "Apodora"; Wikipedia, "Bothrochilus."

18. Coghlan, *Dictionary,* p. 13; Encyclopedia Mythica, http://www.pantheon.org; Wikipedia, "Kiribati."

Chapter 6

1. Wikipedia, "South America."

2. Wikipedia, "Argentina."

3. Wikipedia, "Bolivia."

4. Wikipedia, "Brazil."

5. Wikipedia, "Chile."

6. Wikipedia, "Colombia."
7. Wikipedia, "Ecuador."
8. Wikipedia, "Guyana."
9. Wikipedia, "Paraguay."
10. Wikipedia, "Peru."
11. Wikipedia, "Suriname."
12. Wikipedia, "Uruguay."
13. Wikipedia, "Venezuela."
14. Wikipedia, "Amazon Basin"; Wikipedia, "Manaus"; Wikipedia, "Belém"; Wikipedia, "Iquitos."
15. Wikipedia, "Boa constrictor" and "Boa (genus)."
16. Pope, pp. 14, 156-7; Wikipedia, "Boa constrictor."
17. Heuvelmans, *On the Track,* p. 349.
18. Ibid.
19. Pope, p. 15; Wikipedia, "Anaconda."
20. Eberhart, p. 189; Gould, p. 181; Wikipedia, "Anaconda."
21. Mattison, p. 200; Wikipedia, "List of boine species and subspecies."
22. Gould, p. 174; Mattison, p. 19; Pope, p. 154; Wikipedia, "Anaconda" and "Yellow anaconda."
23. Heuvelmans, *On the Track,* p. 341; Mattison, p. 19; Wikipedia, "Henry Walter Bates."
24. Fawcett, p. 71.
25. Ibid., pp. 65-6.
26. James Tabor, "Journey Men," *Wall Street Journal,* 28 April 2007.
27. Lange, pp. 238-59.
28. Heuvelmans, *On the Track,* p. 344.
29. Ibid., pp. 343-4.
30. Ibid., pp. 345-6.
31. Ibid., p. 346.
32. Ranelagh, pp. 15-16.
33. *Destination Truth,* Sci-Fi Channel, originally aired on 9 April 2008.
34. Pope, p. 228.
35. Eberhart, pp. 189, 467; Rose, p. 386.
36. Heuvelmans, *On the Track,* pp. 349-52.
37. Ibid., pp. 349-50.
38. Ibid., p. 350; Wikipedia, "Capybara."
39. Heuvelmans, *On the Track,* pp. 350-1
40. Ibid., p. 351.
41. Ibid., p. 352.
42. Ibid.
43. Ibid., p. 351.
44. Ibid., pp. 352-3; Eberhart, p. 190.
45. Dash, pp. 194-5.
46. Henry Theiss, "World's biggest snake," Copley News Service, 24 Feb. 1969.
47. Ibid.
48. Wade, p. 35.
49. Ibid.
50. Ibid., pp. 35-6.
51. Ibid., p. 36.
52. Ibid.
53. Ibid.
54. Ibid.
55. Ibid., pp. 36-7.

56. "Boa! Boa! Boa!"; Shuker, "Sachamama."

57. "Boa! Boa! Boa!"; Eberhart, p. 467; Shuker, "Sachamama."

58. *Destination Truth,* Sci-Fi Channel, originally aired on 9 April 2008.

59. Heuvelmans, *On the Track,* p. 356; Wikipedia, "Fritz Müller."

60. Heuvelmans, *On the Track,* pp. 357-8.

61. Ibid., p. 357.

62. Ibid.

63. Ibid., p. 359.

64. Ibid., p. 358.

65. Fawcett, p. 113.

66. Heuvelmans, *On the Track,* pp. 360-1; Wikipedia, "Glyptodon."

67. Wikipedia, "Boa constrictor"; Wikipedia, "Lachesis muta."

68. Quoted in Heuvelmans, *On the Track,* p. 354.

Chapter 7

1. Snakes of North America, http://www.pitt.edu/~mcs2/ erp/SoNA.html; Conant and Collins, pp. 282-415; Stebbins, pp. 171-246; Fauna of Mexico, http://www.vivanatura.org/ AnimalsRept.html.

2. "Florida python invasion: expanded and still growing, UF researcher says," University of Florida News, 15 May, 2008; Arment, *Boss Snakes,* pp. 44-5.

3. Wikipedia, "Horned viper"; Stebbins, p. 229.

4. Charles Moffat, "Sea serpents of Canada," http:// www.lilith-ezine.com/articles/2005/ canadian_seaserpents. html; Rose, pp. 102, 152, 176-7, 212, 250, 362.

5. Moffat, "Sea serpents of Canada"; Arment, *Boss Snakes,* pp. 113-14.

6. Clark, *Unnatural,* pp. 81-3.

7. Arment, *Boss Snakes,* p. 136.

8. Clark, p. 112.

9. Arment, *Boss Snakes,* pp. 169, 172, 174.

10. Arment, *Boss Snakes,* pp. 123, 130, 229; Hall, "Giant snakes," p. 13.

11. Wikipedia, "Great Black Swamp"; Wikipedia, "Limberlost Swamp"; Wikipedia, "Prairie Pothole Region."

12. Arment, "Giant snakes in Pennsylvania"; Wikipedia, "Devil's Den."

13. Arment, *Boss Snakes,* pp. 230, 349.

14. Ibid., p. 230.

15. Ibid.

16. Ibid., pp. 184-5, 283, 317-18; Clark, p. 1.

17. Arment, *Boss Snakes,*, p. 175; Arment, "Giant snakes," pp. 37-8.

18. Arment, *Boss Snakes,* pp. 194-5, 231-4, 339-40.

19. Ibid., pp. 185, 222-3, 310-11, 343; Clark, pp. 1, 115-16; Hall, "More giant snakes," p. 80.

20. Arment, *Boss Snakes,* pp. 88-9, 101-3, 123, 126, 170, 234-9, 284-5, 321-9; Clark, pp. 16, 34-5, 199; Hall, "More giant snakes," p. 87.

21. Arment, *Boss Snakes,* pp. 87-92, 103-5, 124-5, 131, 170-1, 179-80, 182, 186-92, 204, 206-7, 227, 239-46, 285-7, 330, 335-8, 340; Clark, p. 44, 81-3, 320; Eberhart, p. 202; Hall, "More giant snakes," pp. 80-1, 87.

22. Arment, *Boss Snakes,* pp. 92-4, 107-9, 119-21, 125-6, 131-6, 153-9, 176, 207-8, 219-20, 246-9, 278, 287-90, 311, 343-5; Clark, pp. 17, 103-4, 128-9, 134, 151, 173, 255, 257-8; Hall, "More giant snakes," p. 87.

23. Arment, *Boss Snakes,* pp. 93-5, 105-6, 111-12, 126-8, 136-40, 159-63, 171, 182, 196, 208-11, 224-5, 249-55, 279-81, 290 - 5, 341-2, 346-7; Clark, pp. 72, 87, 94, 112, 179-80, 259, 262.

24. Arment, *Boss Snakes,* pp. 95-6, 113-14, 121-2, 140, 163-5, 197, 217-18, 255, 257-9, 297-302, 313, 330-1, 347-8; Clark, p. 266; Hall, "Giant snakes," p. 11.

25. Arment, *Boss Snakes,* pp. 96-7, 114-17, 140-1, 166-7, 176, 193, 201, 213-15, 228, 259-60, 302-7; Clark, pp. 12-15; Hall, "Giant snakes," p. 12.
26. Arment, *Boss Snakes,* pp. 97-8, 141-2, 176, 211-12, 226, 260-1, 266, 313, 319-20; Arment, "Giant snakes," p. 39; "From the past"; Hall, "Giant snakes," pp. 12-13.
27. Arment, *Boss Snakes,* pp. 98-9, 117, 142-3, 167-8, 172-4, 218, 228, 262-6, 307-9, 313-14; Hall, "More giant snakes," p. 87.
28. Arment, *Boss Snakes,* pp. 143-6, 197-200, 202-3, 267-71, 322-4; Hall, "Giant snakes," pp. 14, 17, 26.
29. Arment, *Boss Snakes,* pp. 80-3, 100, 117-18, 146-50, 200, 228, 272-7, 314, 334; Arment, "Giant snakes," p. 39; Hall, "Giant snakes," pp. 19, 27.
30. Arment, *Boss Snakes,* pp. 79, 85, 110, 174, 176-8, 183, 315-16; Arment, "Giant snakes," pp. 39; Hall, "Giant snakes," pp. 28-9.
31. Arment, *Boss Snakes,* pp. 84, 128-9, 316; Coleman, *Mysterious America,* pp. 81-2; "Giant snake rumored to be in Moccasin Creek," *American News* (Aberdeen, SD), 29 August 2001.
32. Wikipedia, "List of rattlesnake species"; Arment, *Boss Snakes,* p. 45.
33. Arment, *Boss Snakes,* p. 349.
34. Ibid.
35. Ibid.
36. Ibid., p. 377.
37. Ibid., pp. 369-70; Clark, p. 115.
38. Arment, *Boss Snakes,* p. 378.
39. Ibid., pp. 357-9.
40. Ibid., p. 360.
41. Ibid., p. 375.
42. Ibid., p. 355.
43. Ibid., pp. 373-4.
44. Ibid., pp. 353-4; Clark, p. 44.
45. Arment, *Boss Snakes,* pp. 351-2.
46. Ibid., pp. 370-1.
47. Ibid., pp. 375-6.
48. Clark, pp. 257-8.
49. Arment, *Boss Snakes,* p. 356.
50. Ibid., p. 361; Clark, p. 87.
51. Newton, *Strange Indiana Monsters,* p. 65.
52. Arment, *Boss Snakes,* pp. 380-1.
53. Ibid., p. 128.
54. Ibid., pp. 361-2.
55. Ibid., p. 379.
56. Ibid., pp. 376-7.
57. Ibid., pp. 362-5.
58. Ibid., p. 352.
59. Ibid., p. 371.
60. Ibid., pp. 365-8.
61. Ibid., p. 372.
62. Ibid., p. 379.
63. Ibid., p. 350.
64. Ibid., pp. 382-4.
65. Ibid., p. 377.
66. Ibid., p. 230.
67. Ibid., pp. 194-5.
68. Ibid., pp. 369-70.

69. Ibid., pp. 357-9.
70. Ibid., p. 123.
71. Ibid., pp. 325-9.
72. Clark, p. 199.
73. Arment, *Boss Snakes,* p. 238.
74. Ibid., p. 239.
75. Ibid., pp. 170-1.
76. Ibid., p. 207.
77. Ibid., 330.
78. Ibid., p. 131.
79. Ibid., p. 375.
80. Ibid., pp. 87-8.
81. Ibid., pp. 179-80.
82. Ibid., p. 88.
83. Ibid., p. 355.
84. Ibid., p. 242.
85. Ibid., pp. 373-4.
86. Ibid., pp. 353-4.
87. Ibid., pp. 103-4.
88. Clark, pp. 81-3.
89. Arment, *Boss Snakes,* p. 246.
90. Ibid., pp. 104-5.
91. Ibid., p. 352.
92. Ibid., pp. 124-5.
93. Ibid., p. 288.
94. Ibid., pp. 287-8.
95. Ibid., p. 278.
96. Ibid., p. 119.
97. Clark, p. 17.
98. Arment, *Boss Snakes,* p. 370-1.
99. Ibid., 288.
100. Ibid., pp. 207-8.
101. Ibid., pp. 134-5.
102. Ibid., p. 289.
103. Clark, p. 255.
104. Arment, *Boss Snakes,* pp. 375-6.
105. Ibid., p. 290.
106. Ibid., pp. 248-9.
107. Clark, pp. 257-8.
108. Ibid., p. 173.
109. Arment, *Boss Snakes,* pp. 105-6.
110. Ibid., p. 356.
111. Ibid., p. 361.
112. Ibid., p. 126.
113. Ibid., pp. 249.
114. Ibid., pp. 224-5.
115. Ibid., pp. 292-4.
116. Ibid., p. 250.
117. Ibid., p. 250-1.
118. Ibid., p. 128.

119. Ibid., pp. 111-12.
120. Ibid., p. 211.
121. Ibid., p. 295.
122. Ibid., p. 255.
123. Ibid., p. 255.
124. Ibid., pp. 256-7.
125. Ibid., pp. 113-14.
126. Ibid., p. 297.
127. Ibid., p. 257.
128. Ibid., p. 258.
129. Ibid., pp. 376-7.
130. Ibid., pp. 121-2.
131. Ibid., pp. 298-9.
132. Ibid., p. 299.
133. Ibid., p. 363.
134. Ibid., p. 301.
135. Ibid., pp. 301-2.
136. Ibid., pp. 259-60.
137. Ibid., pp. 213-15.
138. Ibid., p. 303.
139. Ibid., p. 352.
140. Ibid., p. 201.
141. Ibid., p. 304.
142. Ibid., p. 305.
143. Ibid., p. 176.
144. Ibid., p. 306.
145. Ibid., p. 372.
146. Ibid., pp. 262-6.
147. Ibid., p. 309.
148. Eberhart, p. 200.
149. Arment, *Boss Snakes*, p. 271.
150. Ibid., pp. 117-18.
151. Ibid., pp. 274-6.
152. Eberhart, p. 200.
153. Arment, *Boss Snakes*, p. 377.

Conclusion

1. Arment, *Boss Snakes*, pp. 8, 12-18.
2. Ibid., pp. 12, 148, 174.
3. Ibid., pp. 22-34, 38.
4. Ibid., pp. 34-7; "Captured python said to be world's biggest snake," *The Guardian* (London), 30 Dec. 2003.
5. Arment, *Boss Snakes*, p. 18.
6. Ibid., pp. 19-20.
7. Ibid., p. 19.
8. Ibid., pp. 20-1.
9. Ibid., p. 20.
10. Ibid., 21-2.
11. "Python Snakes, An Invasive Species In Florida, Could Spread To One Third Of US." Science Daily (24 Feb. 2008), http://www.sciencedaily.com/releases/2008/02/080223111456. htm; Arment, *Boss*

Snakes, pp. 66-8.
12. Arment, *Boss Snakes,* p. 181.
13. Ibid., pp. 234-5.
14. Ibid., pp. 187-91.
15. Ibid., p. 176.
16. Ibid., pp. 341-2.
17. Ibid., pp. 161-3.
18. Ibid., pp. 250-1.
19. Ibid., p. 163.
20. Ibid., pp. 111-12.
21. Ibid., pp. 296-7.
22. Ibid., p. 258.
23. Ibid., pp. 140-1.
24. Ibid., p. 201.
25. Ibid., p. 176.
26. Ibid., pp. 97-8.
27. Ibid., pp. 307-8.
28. Ibid., p. 308.
29. Ibid., pp. 202-3.
30. Ibid., pp. 267-70.
31. Ibid., pp. 143-6.
32. Ibid., p. 100.
33. Ibid., pp. 117-18.
34. Ibid., pp. 272-7.
35. Ibid., p. 334.
36. Ibid., pp. 126, 170-1, 283-316, 326-7.
37. Ibid., pp. 74-5.
38. Darren Naish, "Stupidly large snakes, the story so far," Tetrapod Zoology (31 May 2007), http:// scienceblogs. com/tetrapodzoology/2007/05/stupidly_large_snakes_the_stor.php; Alan Turner and Jorge Calvo, "A new sebecosuchian crocodyliform from the Late Cretaceous of Patagonia," *Journal of Vertebrate Paleontology* 25 (March 2005): 87-98.

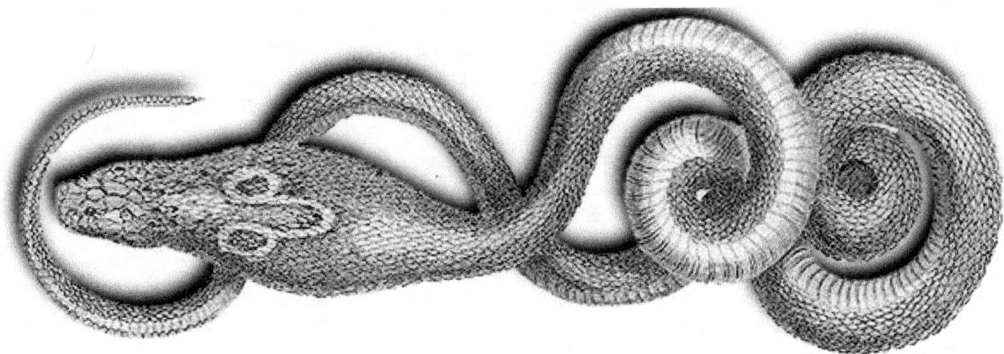

Bibliography

Arment, Chad. *Boss Snakes*. Landisville, PA: Coachwhip, 2008.
- -. "Giant snakes in Pennsylvania." *North American BioFortean Review* 5 (December 2000): 36-43.
    "Boa! Boa! Boa!" *Fortean Times* 104 (November 1997): 18.
Bord, Janet, and Colin Bord. *Unexplained Mysteries of the 20th Century*.
    Chicago: Contemporary Books, 1989.
Clark, Jerome. *Unexplained!* Detroit: Visible Ink, 1999.
- - .*Unnatural Phenomena*. Santa Barbara: ABC-CLIO, 2005.
Coghlan, Ronan. *Cryptosup*. Bangor, No. Ireland: Xiphos, 2005.
- -. *A Dictionary of Cryptozoology*. Bangor, No. Ireland: Xiphos, 2004.
- -. *Further Cryptozoology*. Bangor, No. Ireland: Xiphos, 2007.
Coleman, Loren. *Mysterious America*. New York: Paraview Press, 2001.
Costello, Peter. *In Search of Lake Monsters*. New York: Coward, McCann & Geohegan, 1974.
Dash, Mike. *Borderlands*. New York: Dell, 2000.
Eberhart, George. *Mysterious Creatures*. Santa Barbara: ABC-CLIO, 2002.
    "The fantastic creatures of Angkor,"" Unexplained Earth,
        http://www.unexplainedearth.com/angkor.php.
Fawcett, P.H. *Exploration Fawcett*. London: Hutchinson, 1953.
"From the past: A giant snake." *North American BioFortean Review* 4 (2000): 31.
Freeman, Richard. *Explore Dragons*. Loughborough: Heart of Albion, 2006.
- -. "In the coils of the Naga." http://www.cfz.org.uk/expeditions/00naga/naga1.htm.
Gibbons, William. *Missionaries and Monsters*. Calgary, AB: Creation Generation, 2003.
Gould, Charles. *Mythical Monsters: Fact of Fiction?* London: Studio Editions, 1922.
Hall, Mark. "Giant snakes in the twentieth century." *Wonders* 4 (March 1995): 11-29
- -. "More giant snakes alive!" *Wonders* 4 (September 1995): 80-9.
Heuvelmans, Bernard. *In the Wake of the Sea Serpents*. New York: Hill and Wang, 1968.
- -. *On the Track of Unknown Animals*. London: Kegan Paul, 1995.
Lange, Algot. *In the Amazon Jungle*. New York: G.P. Putnam's Sons, 1912.
Mackal, Roy. *A Living Dinosaur?* Leiden: E.J. Brill, 1987.

Mattison, Chris. *The Encyclopedia of Snakes.* New York: Facts on File, 1995.
Newton, Michael. *Encyclopedia of Cryptozoology.* Jefferson, NC: McFarland, 2005.
- -. *Florida's Unexpected Wildlife.* Gainesville: University Press of Florida, 2007,
- -. *Strange Indiana Monsters.* Atglen, PA: Schiffer, 2006.
Pope, Clifford. *The Giant Snakes.* New York: Alfred A. Knopf, 1980.
Ranelagh, John. *The Agency: The Rise and Decline of the CIA.* New York: Touchstone, 1987.
Rose, Carol. *Giants, Monsters & Dragons.* New York: W.W. Norton, 2000.
Shuker, Karl. *The Beasts That Hide From Man.* New York: Paraview Press, 2003.
- -. *Dragons: A Natural History.* New York: Simon & Schuster, 1995.
- -. *Extraordinary Animals Revisited.* Woolsery, North Devon: CFZ Press, 2007.
- -. "Sachamama: A snake in a shell?" *Fortean Times* 111 (June 1998): 16-17.
- -. *In Search of Prehistoric Survivors.* London: Blandford, 1995.
Twigger, Robert. *Big Snake.* New York: William Morrow, 1999.
Wade, Jeremy, "Snakes alive!" *Fortean Times* 97 (May 1997): 34-8.

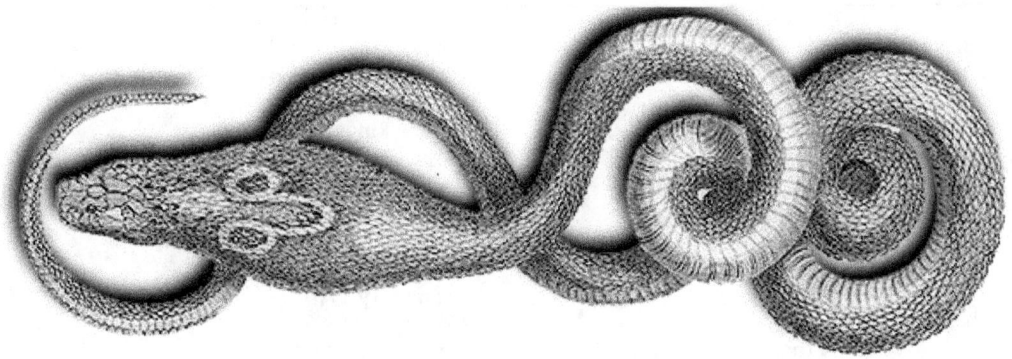

Index

THE CENTRE FOR FORTEAN ZOOLOGY

So, what is the Centre for Fortean Zoology?

We are a non profit-making organisation founded in 1992 with the aim of being a clearing house for information, and coordinating research into mystery animals around the world. We also study out of place animals, rare and aberrant animal behaviour, and Zooform Phenomena; little-understood "things" that appear to be animals, but which are in fact nothing of the sort, and not even alive (at least in the way we understand the term).

Why should I join the Centre for Fortean Zoology?

Not only are we the biggest organisation of our type in the world, but - or so we like to think - we are the best. We are certainly the only truly global Cryptozoological research organisation, and we carry out our investigations using a strictly scientific set of guidelines. We are expanding all the time and looking to recruit new members to help us in our research into mysterious animals and strange creatures across the globe. Why should you join us? Because, if you are genuinely interested in trying to solve the last great mysteries of Mother Nature, there is nobody better than us with whom to do it.

What do I get if I join the Centre for Fortean Zoology?

For £12 a year, you get a four-issue subscription to our journal *Animals & Men*. Each issue contains 60 pages packed with news, articles, letters, research papers, field reports, and even a gossip column! The magazine is A5 in format with a full colour cover. You also have access to one of the world's largest collections of resource material dealing with cryptozoology and allied disciplines, and people from the CFZ membership regularly take part in fieldwork and expeditions around the world.

How is the Centre for Fortean Zoology organized?

The CFZ is managed by a three-man board of trustees, with a non-profit making trust registered with HM Government Stamp Office. The board of trustees is supported by a Permanent Directorate of full and part-time staff, and advised by a Consultancy Board of specialists - many of whom who are world-renowned experts in their particular field. We have regional representatives across the UK, the USA, and many other parts of the world, and are affiliated with other organisations whose aims and protocols mirror our own.

I am new to the subject, and although I am interested I have little practical knowledge. I don't want to feel out of my depth. What should I do?

Don't worry. We were *all* beginners once. You'll find that the people at the CFZ are friendly and approachable. We have a thriving forum on the website which is the hub of an ever-growing electronic community. You will soon find your feet. Many members of the CFZ Permanent Directorate started off as ordinary members, and now work full-time chasing monsters around the world.

I have an idea for a project which isn't on your website. What do I do?

Write to us, e-mail us, or telephone us. The list of future projects on the website is not exhaustive. If you have a good idea for an investigation, please tell us. We may well be able to help.

How do I go on an expedition?

We are always looking for volunteers to join us. If you see a project that interests you, do not hesitate to get in touch with us. Under certain circumstances we can help provide funding for your trip. If you look on the future projects section of the website, you can see some of the projects that we have pencilled in for the next few years.

In 2003 and 2004 we sent three-man expeditions to Sumatra looking for Orang-Pendek - a semi-legendary bipedal ape. The same three went to Mongolia in 2005. All three members started off merely subscribers to the CFZ magazine.

Next time it could be you!

Project Kerinci, Sumatra - 2003
In search of the bipedal ape Orang Pendek

How is the Centre for Fortean Zoology funded?

We have no magic sources of income. All our funds come from donations, membership fees, works that we do for TV, radio or magazines, and sales of our publications and merchandise. We are always looking for corporate sponsorship, and other sources of revenue. If you have any ideas for fund-raising please let us know. However, unlike other cryptozoological organisations in the past, we do not live in an intellectual ivory tower. We are not afraid to get our hands dirty, and furthermore we are not one of those organisations where the membership have to raise money so that a privileged few can go on expensive foreign trips. Our research teams both in the UK and abroad, consist of a mixture of experienced and inexperienced personnel. We are truly a community, and work on the premise that the benefits of CFZ membership are open to all.

What do you do with the data you gather from your investigations and expeditions?

Reports of our investigations are published on our website as soon as they are available. Preliminary reports are posted within days of the project finishing.

Each year we publish a 200 page yearbook containing research papers and expedition reports too long to be printed in the journal. We freely circulate our information to anybody who asks for it.

No. Each year since 2000 we have held our annual convention - the *Weird Weekend* - in Exeter. It is three days of lectures, workshops, and excursions. But most importantly it is a chance for members of the CFZ to meet each other, and to talk with the members of the permanent directorate in a relaxed and informal setting and preferably with a pint of beer in one hand. Since 2006 - the *Weird Weekend* has been bigger and better and held in the idyllic rural location of Woolsery in North Devon. The 2008 event will be held over the weekend 15-17 August.

Since relocating to North Devon in 2005 we have become ever more closely involved with other community organisations, and we hope that this trend will continue. We also work closely with Police Forces across the UK as consultants for animal mutilation cases, and we intend to forge closer links with the coastguard and other community services. We want to work closely with those who regularly travel into the Bristol Channel, so that if the recent trend of exotic animal visitors to our coastal waters continues, we can be out there as soon as possible.

We are building a Visitor's Centre in rural North Devon. This will not be open to the general public, but will provide a museum, a library and an educational resource for our members (currently over 400) across the globe. We are also planning a youth organisation which will involve children and young people in our activities. We work closely with *Tropiquaria* - a small zoo in north Somerset, and have several exciting conservation projects planned.

Apart from having been the only Fortean Zoological organisation in the world to have consistently published material on all aspects of the subject for over a decade, we have achieved the following concrete results:

- Disproved the myth relating to the headless so-called sea-serpent carcass of Durgan beach in Cornwall 1975
- Disproved the story of the 1988 puma skull of Lustleigh Cleave
- Carried out the only in-depth research ever into the mythos of the Cornish Owlman
- Made the first records of a tropical species of lamprey
- Made the first records of a luminous cave gnat larva in Thailand.
- Discovered a possible new species of British mammal - the beech marten.
- In 1994-6 carried out the first archival fortean zoological survey of Hong Kong.
- In the year 2000, CFZ theories where confirmed when an entirely new species of lizard was found resident in Britain.
- Identified the monster of Martin Mere in Lancashire as a giant wels catfish
- Expanded the known range of Armitage's skink in the Gambia by 80%
- Obtained photographic evidence of the remains of Europe's largest known pike
- Carried out the first ever in-depth study of the *ninki-nanka*
- Carried out the first attempt to breed Puerto Rican cave snails in captivity
- Were the first European explorers to visit the `lost valley` in Sumatra
- Published the first ever evidence for a new tribe of pygmies in Guyana
- Published the first evidence for a new species of caiman in Guyana

Other books available from
CFZ PRESS

Other books available from
CFZ PRESS

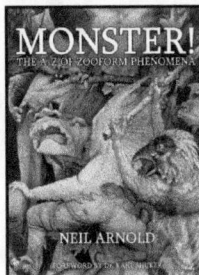

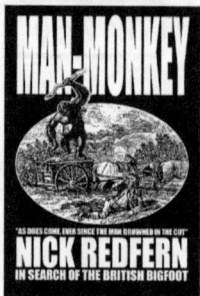

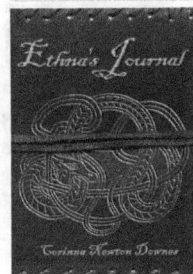

Other books available from
CFZ PRESS

CFZ PRESS

BIG CATS IN BRITAIN YEARBOOK 2008
Edited by Mark Fraser - ISBN 978-1-905723-23-2

£12.50

People from all walks of life encounter mysterious felids on a daily basis, in every nook and cranny of the UK. Most are jet-black, some are white, some are brown; big cats of every description and colour are seen by some unsuspecting person while on his or her daily business. 'Big Cats in Britain' are the largest and most active research group in the British Isles and Ireland. This book contains a run-down of every known big cat sighting in the UK during 2007, together with essays by various luminaries of the British big cat research community.

CFZ EXPEDITION REPORT 2007 - GUYANA
ISBN 978-1-905723-25-6

£12.50

Since 1992, the CFZ has carried out an unparalleled programme of research and investigation all over the world. In November 2007, a five-person team - Richard Freeman, Chris Clarke, Paul Rose, Lisa Dowley and Jon Hare went to Guyana, South America. They went in search of giant anacondas, the bigfoot-like didi, and the terrifying water tiger.

Here, for the first time, is their story...With an introduction by Jonathan Downes and forward by Dr. Karl Shuker.

CENTRE FOR FORTEAN ZOOLOGY 2003 YEARBOOK
Edited by Jonathan Downes and Richard Freeman
ISBN 978 -1-905723-19-5

£12.50

The Centre For Fortean Zoology Yearbook is a collection of papers and essays too long and detailed for publication in the CFZ Journal *Animals & Men*. With contributions from both well-known researchers, and relative newcomers to the field, the Yearbook provides a forum where new theories can be expounded, and work on little-known cryptids discussed.

CENTRE FOR FORTEAN ZOOLOGY 1997 YEARBOOK
Edited by Jonathan Downes and Graham Inglis
ISBN 978 -1-905723-27-0

£12.50

The Centre For Fortean Zoology Yearbook is a collection of papers and essays too long and detailed for publication in the CFZ Journal *Animals & Men*. With contributions from both well-known researchers, and relative newcomers to the field, the Yearbook provides a forum where new theories can be expounded, and work on little-known cryptids discussed.

**CFZ PRESS, MYRTLE COTTAGE,
WOOLFARDISWORTHY BIDEFORD,
NORTH DEVON, EX39 5QR
w w w . c f z . o r g . u k**

Other books available from
CFZ PRESS

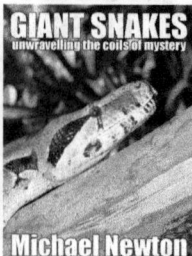

GIANT SNAKES
By Michael Newton
ISBN: 978-1-905723-30-0

£9.99

In this exciting book, Michael Newton takes an overview of the most terrifying uberpredators in the world - giant snakes. Outsized examples of known species as well as putative new species are looked at in detail. From fact to fiction, and from continent to continent, the stories and the science are examined, and the cryptozoological significance of these creatures explained. A fascinating (if slightly scary) book. Well done Michael.

**CFZ PRESS, MYRTLE COTTAGE,
WOOLFARDISWORTHY BIDEFORD,
NORTH DEVON, EX39 5QR**
w w w . c f z . o r g . u k